Biological Processes *in* Living Systems

Biological Processes *in* Living Systems

Toward a Theoretical Biology
Volume 4

C. H. *Waddington,* editor

Routledge
Taylor & Francis Group

LONDON AND NEW YORK

Originally published in 1972 by Transaction Publishers

Published 2017 by Routledge
2 Park Square, Milton Park, Abingdon, Oxon OX14 4RN
711 Third Avenue, New York, NY 10017, USA

Routledge is an imprint of the Taylor & Francis Group, an informa business

Library of Congress Catalog Number: 2011034934

Library of Congress Cataloging-in-Publication Data

Towards a theoretical biology. 4, Essays.
   Biological processes in living systems / C.H. Waddington, editor.
      p. cm. -- (Toward a theoretical biology ; v. 4)
   Originally published: Chicago : Aldine Pub. Co., 1972, as v. 4 of
   Towards a theoretical biology : an IUBS symposium / edited by C.H.
   Waddington.
   Includes index.
   ISBN 978-1-4128-4276-1 (acid-free paper)
      1. Biology--Congresses. 2. Biological systems--Congresses.
   3. Biology--Philosophy--Congresses. I. Waddington, C. H. (Conrad
   Hal), 1905-1975. II. Title.

QH301.T68 2012
570--dc23

2011034934

ISBN 13: 978-1-4128-4276-1 (pbk)

Contents

Preface

This is the fourth volume of papers emanating from a series of Symposia on Theoretical Biology which I have organized on behalf of the International Union of Biological Sciences, at the Rockefeller Foundation's Villa Serbelloni at Bellagio, Italy, during the years since 1966. It will, alas, be the last. Our hosts, whose generosity has allowed itself to be stretched to accommodate us through so many years, have finally decided that we have had our full share of their hard-pressed facilities. It is a tribute to the quite exceptional amenities of the Villa for small, intimate and intense symposia, and to its natural beauties — after all, Pliny the Elder did not choose this site for his Villa, in all Italy, without good reason — that we shall not attempt to prolong this series of meetings in the same form in other surroundings. I am sure that all of those who have taken part in any of our four meetings will join with me in offering especially heartfelt and sincere thanks to the Rockefeller Foundation for its hospitality in the Villa; coupled, perhaps, with the slightly more minatory insistence that they should never allow themselves to forget the quite exceptional facilities they command for encouraging certain sorts of inter-personal communications, the achievement of which is difficult; but finishing on the more benign note of our very best wishes to the newly appointed Director, Dr William C. Olsen, and to Mrs Olsen, who, I respectfully hope, will run the Villa, in their own manner, as worthy successors to Dr John Marshall and his gracious lady, whose personal style as our hosts was an important factor in making the Villa a locale for interactions between whole persons, rather than mere communication-traffic between Theory A and Theory B.

Our first symposium was focussed on the idea that 'the time is ripe to formulate some skeleton of concepts and methods around which Theoretical Biology can grow' to an intellectual and academic stature comparable to Theoretical Physics. We cannot claim to have got the ripening fruit safely into a neat basket, but perhaps we have made some progress. Our offerings have had the titles: Prolegomena, Sketches, Drafts. Now we come forward with a volume entitled Essays — a word which, with all its secondary suggestions of 'try-outs', has a primary meaning of a thought-out and well-organized presentation of a particular (even if not wholly inclusive) point of view. Many of the contributions to this volume seem to me to reach this level.

Further, I think that anyone who looks through all four volumes of the series

Preface

is likely to feel that, by and large, they are tending towards some general point of view about the nature of Theoretical Biology. My invitations went always to the people I thought most likely to have something interesting to say — though inevitably and unfortunately, not all those invited were able to accept — but although there was no conscious selection of people with any particular viewpoint, other than that of an interest in general rather than special problems, something of a focussing on to particular types of question and frameworks of thought seems to be gradually emerging. I do not think that this is sufficiently clearcut to provide a guide to the reader before he tackles the individual essays in this book, but I have made an attempt to formulate my own interpretation of it in the Epilogue.

C. H. WADDINGTON
Edinburgh 1972

The Riemann-Hugoniot catastrophe and van der Waals equation

D. H. Fowler
University of Warwick

This note, an elementary bit of physics, is intended as an accessible illustrative example in René Thom's catastrophe theory announced in the earlier volumes of this series.

▶ *The model.* One of the ways of considering Thom's elementary catastrophes is as follows : Take some system (for example, a fixed amount of liquid and gas) whose local state can be specified by one or more parameters x (for example, density). Then postulate that this state is a minimum x_0 of some, perhaps notional, potential function φ_0 of x. (To appreciate this description it is most instructive, and even essential, to make and play with a Zeeman catastrophe machine [1].) Next, vary the system in any way whatever (for example, by altering its temperature and pressure) and investigate what can happen to the state or, equivalently, to the minimum x_0 of the new φ arising from this perturbation. Under reasonable simplifying hypotheses [2, 3], any nearby φ can be expressed in terms of φ_0 and auxiliary functions $g_1, \ldots, g_k$ by

$$\varphi(x ; u_1, \ldots, u_k) = \varphi_0(x) + u_1 g_1(x) + \ldots + u_k g_k(x) \quad .$$

Thom's paper gives a catalogue of all functions φ_0 for which $k \leqslant 4$.

It should be emphasized at the outset that although these functions φ act precisely as potential functions, it is rarely possible and rarely worthwhile to evaluate them explicitly (as an illustration of this point, the reader is invited to compute the potential function for the catastrophe machine) ; phenomenologically, all that matters is the behaviour around their minima, and this can often be described satisfactorily by Liapunov functions. In fact, in the example that we shall be considering there appears to be no physically relevant potential function.

In this note, we are concerned with the Riemann–Hugoniot catastrophe for which

$$\varphi_0(x) = \tfrac{1}{4} x^4$$

with minimum $x_0 = 0$ (*see* Figure 1). Then any nearby potential function can be expressed as

$$\varphi(x ; u_1, u_2) = \tfrac{1}{4} x^4 + \tfrac{1}{2} u_1 x^2 + u_2 x$$

that is, $g_1(x) = \tfrac{1}{2} x^2$ and $g_2(x) = x$, and only two significant parameters (u_1, u_2) enter into the description (for example, temperature and pressure). Varying

1

Riemann–Hugoniot catastrophe and van der Waals equation

$u=(u_1, u_2)$ can give rise to two different types of potential function : if u lies outside the cusp $4u_1^3+27u_2^2=0$ then there will be one minimum ; if u lies inside, then two (*see* Figures 1 and 2). This situation is expressed by the folded surface of Figure 3, whose equation is

$$x^3+u_1x+u_2=0$$

showing how each point of the u-plane (for example, temperature−pressure plane) has either one or two possible corresponding states (for example, liquid/ gas, or liquid, or gas). The intermediate dotted surface corresponds to maxima of φ and has no significance in the description.

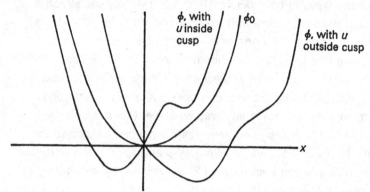

FIGURE 1

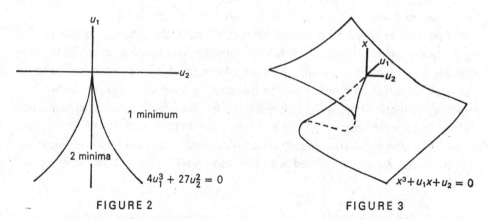

FIGURE 2   FIGURE 3

Summarizing this : each point in the u-plane (Figure 2) gives rise to a potential function $\varphi(x\,;u)$ (Figure 1) and every potential function close to φ_0 (in some appropriate sense) arises in this way ; the state x of the system represented by the potential function $\varphi(x\,;u)$ will be at a minimum of $\varphi(x\,;u)$; so that the location of the minima, as u varies, gives the state surface of the process (Figure 3). The general mathematical theory of relationship between φ_0 and φ is only now being developed ; details of it, and very many other applications of catastrophe theory will be given in the paper by Zeeman [1].

Inside the curve $4u_1^3 + 27u_2^2 = 0$ it is necessary to specify to which of the two minima the state corresponds. Whatever rule governs this choice, there must be a discontinuity somewhere in the state on each line $u_1 = c$ with $c \leqslant 0$. This simple mechanism for describing a very general class of discontinuities is one of the remarkable features of Thom's model.

Several different rules governing the choice of minimum will be discussed later.

▶ *The liquid/gas critical point.* The parenthetic observations above suggest that the known behaviour of the liquid/gas critical point might be a good candidate on which to try out the Riemann–Hugoniot description. This feeling is reinforced by the mere appearance of the isotherms, shown in Figure 4. (Three isotherms are shown, at critical, sub-critical, and super-critical temperatures ; the sub-critical has added portions corresponding to super-heating and -cooling.) Regarding pressure and temperature as the control variables, and volume as the state variable, drawing the state surface over the pressure/temperature plane, and filling in the 'van der Waals loops' gives Figure 5, which has the same appearance as Figure 4 in the neighbourhood of the critical point.

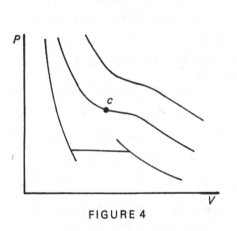

FIGURE 4

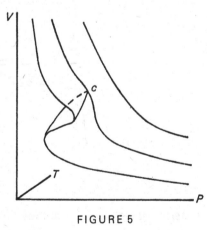

FIGURE 5

What I shall do now is work out three simple exercises in translation between an equation of state and the equation

$$x^3 + u_1 x + u_2 = 0$$

of the surface of Figure 3.

Riemann–Hugoniot catastrophe and van der Waals equation

1. Van der Waals equation. Starting with van der Waals equation,

$$\left(P+\frac{a}{V^2}\right)(V-b)=rT$$

we evaluate the critical point $(P_c, V_c, T_c)=\left(\frac{a}{27b^2}, 3b, \frac{8a}{27br}\right)$ and then normalize, by choosing a, b, and r suitably, to make this point $(1, 1, 1)$. This gives the absolute van der Waals equation for Figure 5 :

$$\left(P_a+\frac{3}{V_a^2}\right)(V_a-\tfrac{1}{3})=\tfrac{8}{3}T_a \ .$$

Next write $V_a=1/X_a$, where X_a is the absolute density (which, incidentally, is much more sensible as a local parameter than the volume). Finally, shift coordinates to the critical point $(1, 1, 1)$, to correspond to the description of Figure 3 : then writing $p=P_a-1$, $x=X_a-1$, $t=T_a-1$ gives

$$x^3+\tfrac{1}{3}(8t+p)x+\tfrac{1}{3}(8t-2p)=0 \ ,$$

that is, the standard equation

$$x^3+u_1x+u_2=0$$

with $u_1=\tfrac{1}{3}(8t+p)$ and $u_2=\tfrac{1}{3}(8t-2p)$.

Hence van der Waals equation is precisely our standard equation referred to the oblique axes in the (p, t) plane, shown in Figure 6, with the folded surface over the shaded cusp region.

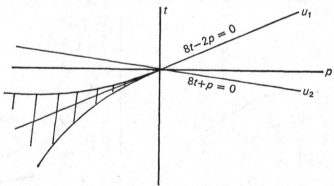

FIGURE 6

2. Berthelot's equation. Another well-known equation of state is Berthelot's equation

$$\left(P+\frac{a}{TV^2}\right)(V-b)=T \ .$$

Following exactly the same procedure gives this equation

$$x^3+\tfrac{1}{3}(pt+p+9t)x-\tfrac{2}{3}(pt-p+3t)=0 \ .$$

4

the standard equation referred to the curvilinear axes in the (p, t) plane of Figure 7.

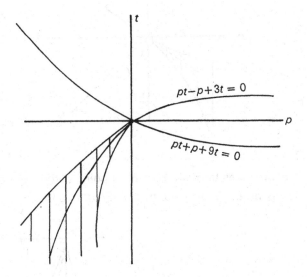

FIGURE 7

3. The reverse procedure. Starting with the (physically meaningless) hypotheses that it is equally effective (in absolute coordinates) to boil a liquid either by raising the temperature, or by lowering the pressure ; and that keeping $p+t$ constant will maintain the matter in the same state (see Figure 8), we might be led to consider an equation of state of the form

$$x^3 + \alpha(t+p)x + \beta(t-p) = 0$$

in which the cusp is symmetrically placed in the $t \leqslant 0$, $p \leqslant 0$ quadrant. Reversing the previous procedure, and making the choice $\alpha = \beta = 3/2$ (which results in the simplest and most familiar-looking equation) we get, in absolute coordinates,

$$\left(P_a + \frac{1}{V_a{}^2}\right)(V_a - \tfrac{1}{2}) = \tfrac{1}{2}T_a - \frac{1}{3V_a{}^2} \quad .$$

▶ *Maxwell's convention, and some others.* Some procedure is required for specifying the choice between the two minima within the cusp region. The classical argument due to Maxwell is that, for thermodynamical reasons, the van der Waals loops should be cut off so as to make the shaded areas of Figure 9 equal. This almost corresponds to saying that the lowest of the two minima is chosen, and it is this latter that Thom has called the *Maxwell convention*. The equation of the coexistence curve, shown dotted in Figure 9, with respect to

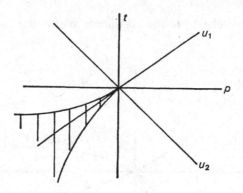

FIGURE 8

absolute (p, x) coordinates through the critical point, is then $x^2+p=0$, and the discontinuity occurs on the line $u_2=0$, $u_1 \leqslant 0$ (Figure 10).

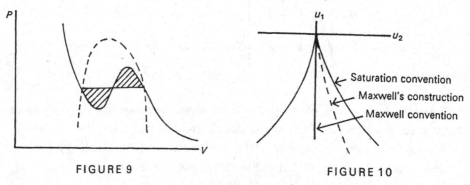

FIGURE 9   FIGURE 10

The construction described by Maxwell actually corresponds to locating the discontinuity on another line (whose precise equation I have been unable to compute) emanating from the origin and lying with the cusp and, to illustrate the effect of such a perturbation, consider a simple case called the *saturation convention* that one minimum is always preferred as soon as it comes into existence. (Such a convention might be conceived as operating within a small neighbourhood of the critical point, whilst outside this neighbourhood it could shade off into another, for example, Maxwell's convention.) Supposing that the positive minimum is always preferred, corresponding to the discontinuity on the right hand edge of the cusp, with a coexistence curve given by $x=\theta$, $u_1=-3\theta^2$, $u_2=2\theta^3$ for $\theta \geqslant 0$, and $x=-2\theta$, $u_1=-12\theta^2$, $u_2=16\theta^3$ for $\theta \leqslant 0$. In the case of van der Waals equation this will give rise to the coexistence curve in the (p, x) plane shown in Figure 11, with equation $p+12x^2-16x^3=0$ if $x \leqslant 0$, and

6

$p + 3x^2 + 2x^3 = 0$ if $x \geqslant 0$. This shows the typical 'non-analytic' behaviour at the critical point, together with a cubic term not present when using Maxwell's convention.

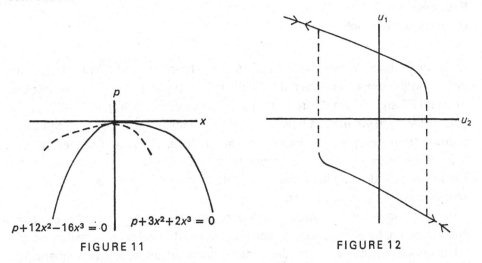

$p + 12x^2 - 16x^3 = 0$ $p + 3x^2 + 2x^3 = 0$

FIGURE 11   FIGURE 12

A final convention, which does not apply to the liquid/gas critical point but does have some bearing on the ferromagnetic critical point, and which is the appropriate convention for the catastrophe machine, is the *delay convention* : the system will remain in a minimum until this minimum disappears. This will result in a hysteresis phenomenon : moving across the cusp on the line $u_1 = c$, $c \leqslant 0$ will give rise to a typical state diagram of Figure 12.

References

1. E. C. Zeeman, Differential equations for the heartbeat and nerve impulse, in (C. H. Waddington, ed.) *Towards a Theoretical Biology 4 : Essays.* pp. 8-67 ; 276-82 (Edinburgh University Press 1972).

2. R. Thom, Une théorie dynamique de la morphogénèse, in (C. H. Waddington, ed.) *Towards a Theoretical Biology 1 : Prolegomena.* pp. 152-65 (Edinburgh University Press 1968).

3. R. Thom, Topological models in biology, in (C. H. Waddington, ed.) *Towards a Theoretical Biology 3 : Drafts,* pp. 89-116 (Edinburgh University Press 1970).

Differential equations for the heartbeat and nerve impulse

E. C. Zeeman
University of Warwick

We abstract the main dynamical qualities of the heartbeat and nerve impulse, and then build the simplest mathematical models with these qualities. The results are flows on R^2 and R^3, which are related to Thom's cusp catastrophe [1–3]. In the case of the heart we explain how many of the complexities of the beat can be deduced from a simple behaviour of the muscle fibre. In the case of the nerve impulse we fit the model to the experimental data of Hodgkin and Huxley [4, 5]. The model provides an alternative to the latter's own equations, and suggests alternative underlying chemistry.

In these two examples catastrophe theory provides not only a better conceptual understanding, by giving a single global picture that enables an overall grasp of the phenomenon, but also provides explicit equations for testing experimentally. The novelty of the approach lies in modelling the *dynamics* (which is relatively simple) rather than the *biochemistry* (which is relatively complicated). This approach might be useful for a large variety of phenomena in biology, whenever there is a trigger mechanism leading to some specific action. In an appendix we suggest how it might be applied to evolution.

Mathematically the equations that we derive are interesting, because they are generalizations of the Van der Pol and Liénard equations.

Part One

1.1. THREE QUALITIES

The three dynamic qualities displayed by heart muscle fibres and nerve axons are:
(1) stable equilibrium;
(2) threshold, for triggering an action;
(3) return to equilibrium.
The third quality can be divided into two cases according as to whether or not the return is smooth:
(3a) jump return (heart);
(3b) smooth return (nerve).
In the first half of the paper our objective will be to take these three qualities as 'axioms', and derive mathematical models by means of qualitative mathematical

8

argument. We state the models in Theorems 1, 2 in sections 1.7 and 1.10 respectively, and apply them *qualitatively* to heart and nerve in sections 1.8 and 1.11. In order to keep the mathematics as translucent as possible, we only touch upon the physiology where necessary.

In the second half of the paper the models are developed *quantitatively* into a form that could be used for prediction, and for this we need to go more deeply into the physiology and numerical data. In sections 2.1 to 2.3 we discuss the heart, and in sections 2.4 to 2.10 the nerve, concluding both cases by suggesting how the models might be used in experiments. But first we must explain why we chose these three qualities.

There are two states to the heart, the *diastole* which is the relaxed state, and the *systole* which is the contracted state. If the heart stops beating it stays relaxed in *diastole*, which is therefore the stable equilibrium of Quality 1. What makes the heart contract into systole is a global electrochemical wave emanating from a pacemaker. As the wave reaches each individual muscle fibre, it triggers off the action of Quality 2, that makes the fibre rapidly contract. Each fibre remains contracted in the systole state for about 1/5 second, and then rapidly relaxes again—in other words the fibre obeys the jump return to equilibrium of Quality 3a. Therefore the three qualities describe the *local* behaviour for an individual muscle fibre, and our first objective is to describe this by an ordinary differential equation, which we achieve in Theorem 1. In section 2.1 we deal with the *global* structure of the heart and its four chambers, and in section 2.2 describe the pacemaker wave by a separate geodesic flow.

Now for the nerve impulse. A *neuron*, or nerve cell is in some ways like a tree (*see* Figure 1).

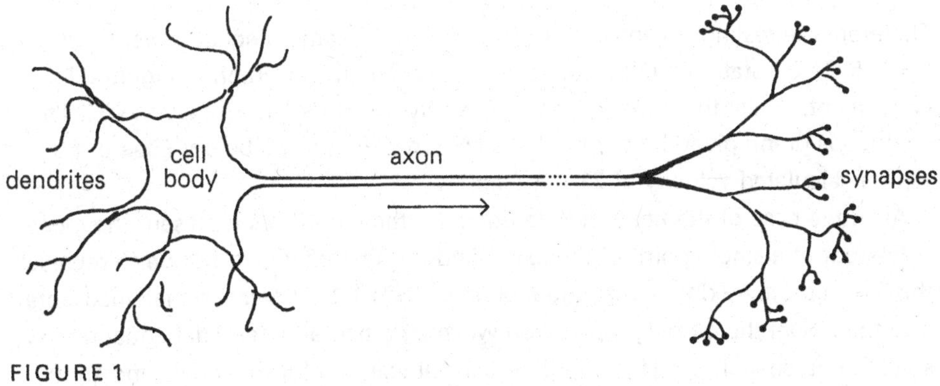

FIGURE 1

9

Heartbeat and nerve impulse

Corresponding to the roots of the tree are the *dendrites*, through which the neuron receives messages. Out of the cell body grows the *axon*, which is like the trunk of the tree. Just as the trunk of a tree divides into many branches, ending in thousands of leaves, so the axon divides into many branches, ending in thousands of *synapses*, which touch the dendrites of other neurons. It is the axons that connect up the whole nervous system, and comprise the white matter of our brains, while the dendrites and cell bodies comprise the grey matter. Messages travel along the axon to the synapses, and are thus transmitted to other neurons. This is the basic mechanism underlying the working of the brain, and so it is interesting to examine the nature of the message. A message consists of a series of *spikes* (*see* Figure 2).

Each spike is an electrochemical phenomenon that lasts about 1 msec ($=10^{-3}$ sec), and travels very fast along the axon. The velocity is between 10 and 100 metres per second, depending upon the type of nerve. The simplest way to observe a spike is to put electrodes inside and outside the axon in order to measure the potential difference v of the inside relative to the outside, and then plot v against time (*see* Figure 3). While no message is passing the membrane surrounding the axon is polarized, and v remains constant about -65 mV (1 mV $=10^{-3}$ volts). This constant is called the *resting potential*, and represents the stable equilibrium of Quality 1. As the spike moves along the axon it triggers off a rapid depolarization of the membrane, called the *action potential*, which causes v to swing to $+40$ mV, representing the action of Quality 2. When the spike has passed, the membrane repolarizes relatively slowly, representing the smooth return to equilibrium of Quality 3b. These features can be seen in Figure 3, which is taken from Hodgkin [5, p. 64]. Notice that the potential V shown in Figure 3 is relative to the resting potential,

$$V = v + 65 \quad .$$

Therefore the resting potential is $V=0$, while the maximum action potential is $V=115$. This notation will be convenient later. The lower graph in Figure 3 is experiment, whilst the upper graph is given by the Hodgkin–Huxley model. The analogous graph given by our model is shown in Figure 26 below (*see* p. 60), with a calculated velocity of 21·9 m/sec.

As in the case of the heart muscle fibre, the three qualities represent the *local* behaviour at a single point of the axon, and our first objective will be to represent this by an ordinary differential equation in section 1.11. Later we shall incorporate into the model the *global* propagation wave by a partial differential equation ; we shall find in section 2.8 Theorem 3 an explicit solution for the wave, and show

10

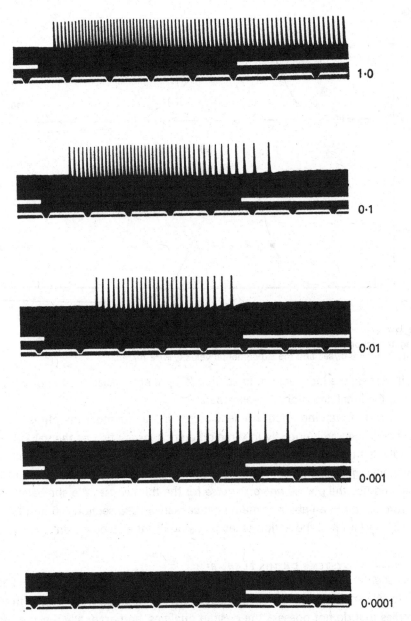

FIGURE 2
Impulses set up in optic nerve fibre of *Lumulus* by one second flash of light with relative
intensities shown at right. The lower white line marks 0·2 second intervals and the gap in the
upper white line gives the period for which the eye was illuminated. (*after* Hartline, 1934)

Heartbeat and nerve impulse

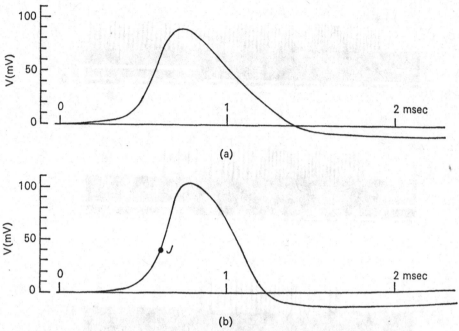

FIGURE 3
Propagated action potentials in (a) theoretical model, and (b) squid axon at 18·5°C. The calculated velocity was 18·8 m/sec and the experimental velocity 21·2 m/sec.

how it triggers the local action. In section 2.4 we emphasize the importance of keeping the local and global ideas separate.

One word of warning about nomenclature : in the literature the phrase *action potential* is used ambiguously in three different senses : first as the value $V=115$, secondly as the graph pictured in Figure 3, and thirdly as incorporating all the accompanying biochemical events as well. We shall tend to use it in the second sense, and use the phrase *nerve impulse* for the third sense. We shall also use the word *action* in an abstract mathematical sense (*see* section 1.6 and Figures 8 and 13). In Figure 3 the action takes place during the steep ascent.

1.2. DYNAMICAL SYSTEMS ON THE PLANE R^2

In this section we shall start from the three qualities, and derive mathematical models. We shall proceed by giving a sequence of examples, starting with trivial examples that do not possess the desired qualities, and gradually making them more complicated step by step, until they do. We shall show that each step is necessary in order to achieve the qualities, and in this way we shall arrive at the simplest model.

First we try and represent the qualities by a differential equation in the Euclidean plane, R^2. (The words *dynamical system, differential equation, flow, vector field* all mean the same thing.) We shall show eventually that using R^2 implies 3a, and for 3b it will be necessary to use R^3. However in this section we confine ourselves to R^2. Quality 1 requires a fixed point, which we can take to be the origin $0 \in R^2$. Let

$$\dot{v} = Av$$

be the linear approximation to the differential equation at 0, where A is a real 2×2 matrix. Since the equilibrium is stable, the two eigenvalues of A each have negative real part. We shall now give an argument to show that one has large negative real part, and the other small negative real part, implying that they must both be real.

Consider Quality 2 : what is an *action* ? It is something that is noticeable, and therefore must involve a reasonably large change in one of the coordinates, and take a reasonably short time. Therefore an action must be represented by large vectors of the vector field, or in other words by part of an orbit along which the flow is fast. More briefly we shall call this a *fast orbit*. Quality 2 implies there is a fast orbit near 0, and the presence of large vectors near 0 means that one of the eigenvalues must have large modulus. We call this the *fast eigenvalue*. Meanwhile it must be large in comparison with something else, and so the other eigenvalue must have relatively small modulus ; we call this the *slow eigenvalue*.

▶ *Example 1. Trivial linear case.* Let x, b be coordinates in R^2. Admittedly it is unusual to use letters at opposite ends of the alphabet, but this will emphasize the difference between fast and slow, and will be later useful notation for the cusp catastrophe (section 1.10). Choose a small positive constant ε.

The equations are

$$\varepsilon \dot{x} = -x$$
$$\dot{b} = -b \quad .$$

Therefore the matrix

$$A = \begin{pmatrix} -\dfrac{1}{\varepsilon} & 0 \\ 0 & -1 \end{pmatrix}$$

The fast eigenvalue is $-\dfrac{1}{\varepsilon}$, and we call $\varepsilon \dot{x} = -x$ the *fast equation*, because x decays exponentially fast by $e^{-t/\varepsilon}$. The slow eigenvalue is -1, and we call $\dot{b} = -b$ the slow equation, because b decays more slowly by e^{-t}. The orbits are shown

Heartbeat and nerve impulse

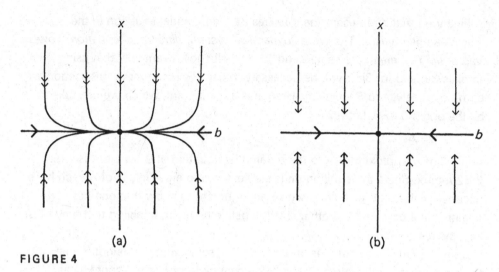

FIGURE 4

in Figure 4a, but we shall find it more convenient to represent the flow symbolically by Figure 4b, for reasons which we shall now explain.

1.3. THE SLOW MANIFOLD AND THE FAST FOLIATION

In all the examples we shall give, the fast equation will always be of the form $\varepsilon\dot{x} = -f(x, b)$, where $f(x, b)$ is a function of x, b that vanishes at the origin. We define the *slow manifold* by putting $\varepsilon = 0$; in other words the slow manifold is the curve through the origin given by $f(x, b) = 0$. In Example 1 the slow manifold is given by $x = 0$, namely the b-axis. In our diagrams we shall always indicate the slow manifold by a thick line with single arrows. Meanwhile define the *fast foliation* to be the family of lines parallel to the x-axis. Except in the neighbourhood of the slow manifold the vector field is approximately parallel to the fast foliation, because the x-component dominates the b-component. Therefore in our diagram we shall always indicate the fast foliation by double arrows parallel to the x-axis, oriented towards x positive where $f < 0$, and oriented in the opposite direction where $f > 0$. If equilibrium is disturbed, we can approximately describe the way the system returns to equilibrium as follows : *the system first homes rapidly down the fast foliation towards the slow manifold, and then homes relatively slowly along the slow manifold back to equilibrium.*

As $\varepsilon \to 0$ the fast foliation speeds up, and the system homes more rapidly towards the slow manifold. In the limit $\varepsilon = 0$, the system jumps instantaneously onto the slow manifold. Therefore, in effect, the system is confined to the slow

14

manifold, where its behaviour is governed by the slow equation. Summarizing we can say :

(1) the fast foliation is parallel to the x-axis ;

(2) the fast equation determines the slow manifold ;

(3) the slow equation determines behaviour on the slow manifold.

We emphasize that these statements are precise when $\varepsilon=0$, and first order approximations in ε when $\varepsilon\neq0$. They are justified by the following lemma.

▶ *Lemma 1*

The tangents at the origin to the fast foliation and the slow manifold are, to first order in ε, the eigenspaces of the fast and slow eigenvalues, respectively.

Corollary 1. The slow manifold is transversal to the fast foliation in the neighbourhood of the origin.

Corollary 2. The slow manifold is an approximation of the invariant submanifold determined by the slow eigenvalue.

Proof. The matrix for the linear approximation of the differential equation at the origin is

$$A = \begin{pmatrix} -\dfrac{a_{11}}{\varepsilon} & -\dfrac{a_{12}}{\varepsilon} \\ a_{21} & a_{22} \end{pmatrix}$$

where $f(x, b) = a_{11}x + a_{12}b + $ higher terms,

$$b = a_{21}x + a_{22}b + \text{higher terms},$$

and each $|a_{ij}|$ is small compared with $\dfrac{1}{\varepsilon}$. Since both eigenvalues are negative, one large and one small, the trace of A must be large negative, and so $a_{11}>0$. Hence to first order in ε, the fast and slow eigenvalues are $\dfrac{a_{11}}{\varepsilon}$ and $\dfrac{a_{11}a_{22}-a_{12}a_{21}}{a_{11}}$ respectively, with corresponding eigenvectors $(1, 0)$ and $(a_{12}, -a_{11})$ which are tangents to the fast foliation and slow manifold, as required. Corollary 1 follows from the fact that $a_{11}\neq0$, and Corollary 2 from the lemma.

We now illustrate these ideas with some more examples.

▶ *Example 2*

$$\varepsilon\dot{x} = -(x+b)$$
$$\dot{b} = -b$$

Here the slow manifold is $x+b=0$. The reader can verify that the precise invariant submanifold determined by the eigenvalue -1 is in fact the line $(1-\varepsilon)x+b=0$, to which the slow manifold is an approximation.

15

Heartbeat and nerve impulse

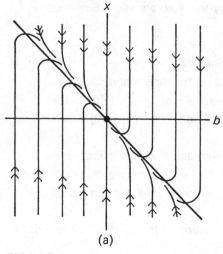 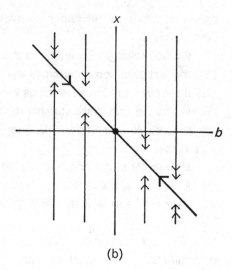

(a)   (b)

FIGURE 5

▶ *Example 3*

$$\varepsilon x = -(x+b)$$
$$b = x$$

The change from Example 2 to Example 3 is slight but subtle. The slow manifold is the same because the fast equation is unchanged. The behaviour on the slow maniiold is the same, because on the slow manifold $-b=x$, and so the slow equation, which determines the behaviour, is in effect the same. Thus they have the same picture, Figure 5. When we reach Example 6 we shall explain why the change was mathematically necessary at this stage in order to later achieve Quality 3. (*See* Remark after Theorem 1 in section 1.7.) Biologically that change has subtle implications, and it is worthwhile digressing for a moment to explain why.

1.4. BIOLOGICAL DIGRESSION

When a cell develops in the embryo its initial task is to achieve an identity, as a nerve cell, or liver cell, or whatever. Unless it does so, the species will become unstable and die out. This implies a very strong homeostasis. Therefore if we model the underlying chemistry by a dynamical system on R^n (with n perhaps very large), the equilibrium point must be very stable, and so all the eigenvalues have large negative real part. Once a cell has achieved its identity, its next task is to develop its dynamic, specific to its job : a nerve cell must conduct impulses,

16

a liver cell must manufacture glucose, and so on. To do this the cell must weaken its equilibrium, so that some of the internal chemistry can vary according to a very specific sequence of reactions, while the rest of the metabolism is directed at holding the identity in a tight homeostasis. Mathematically this is represented in the dynamical system by a slowing down of one eigenvalue, allowing the system to vary along a specific 1-dimensional slow manifold M. Meanwhile the homeostasis is represented by a fast foliation of $(n-1)$-dimensional disks transversal to M, which rapidly damps down any perturbations off M. For instance we could represent this situation by adjoining to Examples 1, 2, or 3 another $n-2$ fast equations of the form $\varepsilon \dot{y} = -y$.

Suppose now that each coordinate represents the concentration of some enzyme, or the productivity of some enzyme-system (relative to equilibrium). Then Example 1 describes a metabolism in which each enzyme-system is an independent self-correcting mechanism for stabilizing its concentration. In Example 2 the enzyme-systems are no longer independent, because, although b is still slowly self-correcting, any perturbation of b causes a rapid change of x, and x only returns slowly to normal with b. In Example 3 the subtle difference is that the two enzyme-systems become linked : in other words b is no longer self-correcting, but a perturbation of b induces a rapid response in x, which is then the mechanism for restabilizing b. The mathematics will show that it is necessary to develop a linkage between 2 or 3 enzyme-systems before a cell can develop the three desired qualities. The same remarks apply not only to development, but also to evolution (*see* section 2.11). This ends the digression.

1.5. NON-LINEAR EXAMPLES ON R^2

▶ *Example 4*

$$\varepsilon \dot{x} = -(x^3 + x + b)$$
$$\dot{b} = x$$

The interest of this example is the non-linearity. At the origin Example 4 has Example 3 as its linear approximation, and so the local behaviour is as before. Globally the slow manifold is curved, and away from the origin the situation is not unlike Example 1. From the point of view of the orbit structure, all the first four examples are diffeomorphic. Figure 6a shows the orbit structure, and Figure 6b the symbolic diagram.

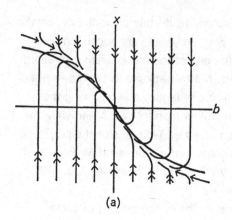

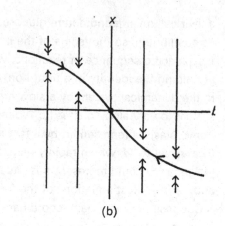

(a)

(b)

FIGURE 6

▶ *Example 5.* Liénard and Van der Pol.

$$\varepsilon\dot{x} = -(x^3 - x + b)$$
$$b = x$$

Example 5 is obtained from Example 4 by a change of sign in one of the terms of the fast equation. This has the effect of bending the slow manifold into the S-shaped cubic curve, as shown in Figure 7. In so doing we have lost Quality 1, because the origin has switched from being an attractor to a repellor (from being a sink to a source). However, mathematically the example has several points of interest. Not only has the origin switched, but, from the point of view of the foliation, the entire middle piece of the slow manifold has switched from being an attractor to a repellor—that is why we have indicated it by a dashed line in Figure 7b. At the same time an attracting closed orbit (or limit cycle) has appeared, as shown in Figure 7a. The flow is diffeomorphic to that of the Van der Pol equation. In fact if we eliminate b by differentiating the fast equation and substituting the slow equation, we obtain a single second-order equation in x

$$\varepsilon\ddot{x} + (3x^2 + 1)\dot{x} + x = 0 \quad,$$

which is an example of the Liénard equation and differs only from the Van der Pol equation by a change of coefficients.

The points T, T' where the fast foliation is tangent to the slow manifold are given by $3x^2 = 1$. Let M' denote the attracting part of the slow manifold, which is given by the equation and inequality

$$x^3 - x + b = 0, \qquad 3x^2 \geqslant 1 \quad.$$

18

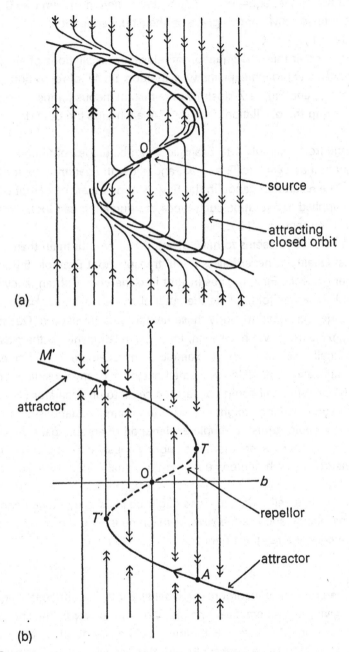

(a)

(b)

FIGURE 7

Heartbeat and nerve impulse

If we speed up the fast equation by letting $\varepsilon \to 0$, then in the limit $\varepsilon = 0$ the system is confined to M', and is given by the single equation

$$(3x^2 - 1)\dot{x} + x = 0$$

which is none other than the equation of relaxation oscillations of electrical engineering. This approach explains why the domain M' of relaxation oscillations is disconnected, and Figure 7 illustrates the instantaneous jumps $T \to A$, $T' \to A'$ that occur during the oscillation cycle $TAT'A'$ each time the boundary of M' is reached.

The change from Example 4 to Example 5 can be made continuous, for write the fast equation as $\varepsilon \dot{x} = -(x^3 + ax + b)$, and allow the parameter a to run from $+1$ to -1. The resulting change in the flow is none other than Hopf bifurcation [6]. When applied to evolution such types of bifurcation can look remarkably Lamarckian (*see* section 2.11).

Although Example 5 seems to have diverted us from our main theme, because we have lost Quality 1, nevertheless it will transpire that Example 5 is useful in several other respects. First, it gives insight into the points of tangency of fast foliation and slow manifold, where the latter changes quality from attractor to repellor; we shall need to study these next in order to capture Quality 2. Secondly it explains, as we have seen, the close relationship to the classical equations. Finally the closeness of Example 5 and Example 6 has interesting biological implications, as follows. A small chemical change inside a heart cell (Example 6) or nerve cell could convert the cell into an oscillator (Example 5). In particular such a change might occur during normal metabolism if the cell is not used. This would furnish a simple explanation of ectopic pacemakers in the heart (secondary pacemakers that arise spontaneously if the main one fails) and self-firing neurons (such neurons occur in the visual cortex – *see* section 2.10, paragraph 7).

In skeletal muscle cells such changes might be caused by fatigue and stress, and cause oscillation such as the sewing-machine legs of a climber, or the muscular tremors of a religious dancer.

1.6. THRESHOLD

We tackle the problem of incorporating Quality 2 into the differential equation. This means giving a mathematical interpretation of the words 'threshold, for triggering an action'. We have already interpreted *action* to mean a piece of fast orbit, sufficiently long to be noticeable, and this has given rise to the discussion of fast and slow. What is a *trigger*? We suggest that a trigger is a specific

20

perturbation away from the equilibrium position, imposed by an external agent (external to the differential equation under discussion). We have already considered the local behaviour near the origin : given a small perturbation the system returns by fast orbit to the slow manifold, and then returns slowly to equilibrium. Such local behaviour can hardly be called an action, because the length of fast orbit involved is less than the size of the trigger, whereas the words 'trigger' and 'action' usually carry the overtone that the action is noticeably larger than the trigger. Certainly this is the case for both heart and nerve.

Now the slow manifold is transversal to the fast foliation near the origin, by Corollary 1 to Lemma 1, and this local behaviour persists as long as it remains transversal. Therefore the non-local behaviour required by Quality 2 implies the existence of a point T of non-transversility, where the fast foliation is tangent to the slow manifold. We define this point to be the *threshold*. We can assume the tangency at T is generic, that is to say quadratic (as in Figures 7a and 8), because otherwise, if it were cubic or higher, this would be a non-generic situation, unstable, and removable by an arbitrarily small perturbation of the differential equation. Since the tangency at T is generic, then by continuity of the fast foliation, the slow manifold must change from attractor to repellor at T. Summarizing, we have shown :

▶ *Lemma 2*

A dynamical system in R^2 possessing Qualities 1 and 2 must have the qualitative picture shown in Figure 8 near the equilibrium position.

▶ *Remark 1*. Notice that Figure 8 differs from Figure 7b in the orientation of the flow on the slow manifold near T.

▶ *Remark 2*. Why not bend the fast foliation rather than the slow manifold ? Differentiably it would be equivalent, but there is a biological reason for not doing so. To bend the fast foliation would need large first partial derivatives of the vector field, and ultimately these are bounded by the energy available in the underlying chemistry. On the other hand to bend the slow manifold requires comparatively little energy. It is the energy bound on the partial derivatives that makes the fast orbits approximately parallel, and justifies extending the fast foliation outside the immediate neighbourhood of 0.

1.7. SOLUTION IN THE CASE OF THE JUMP RETURN

Finally we must elaborate Figure 8 so as to incorporate Quality 3, the return to equilibrium. The only way to stop a fast orbit is to catch it at the bottom point A with an attracting piece of slow manifold. Then the only way to achieve a smooth

Heartbeat and nerve impulse

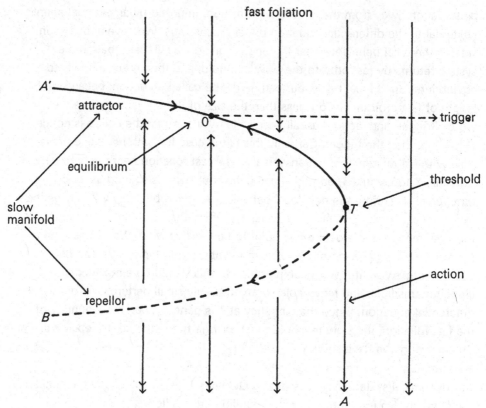

FIGURE 8

return would be to glue this piece of slow manifold back onto the loose end at
A'. However this is impossible because in so doing we should have to bend the
slow manifold round and unwittingly create another threshold point T', where,
not only would the slow manifold switch from attractor to repellor and be
impossible to glue onto A', but also we should shoot off onto another piece of
fast orbit $T'A'$, disrupting the smooth return. Therefore we have to make a jump
return, and :

▶ *Lemma 3*

In R^2 a smooth return is impossible.

Having admitted that another threshold T' is inevitable, the simplest way of
constructing a slow manifold with all the desired properties is to glue the resulting new piece of repellor onto the loose end B. The simplest algebraic equation
of such a curve is the cubic that we used in Example 5 and Figure 7, namely

$x^3 - x + b = 0$. The only difference from Example 5 is the behaviour on the slow manifold, which is determined by the slow equation. All we need to do is to move the equilibrium point from the repellor part to the attracting part of the slow manifold, that is to say from $x=0$ to $x=x_0$, say, where $x_0 > \dfrac{1}{\sqrt{3}}$.

Summarizing, we have shown :

▶ *Theorem 1*

There exists a dynamical system on R^2 possessing Quality 1, 2, and 3a. The simplest example is the following :

▶ *Example 6.* (Heartbeat.)

$$\varepsilon\dot{x} = -(x^3 - x + b)$$
$$b = x - x_0$$

where x_0 is a constant greater than $\dfrac{1}{\sqrt{3}}$.

The orbit structure and symbolic diagram are shown in Figure 9, which can be compared with Figure 7.

▶ *Remark.* Let (x_0, b_0) denote the equilibrium point. Consider the effect of replacing the slow equation by $\dot{b} = -(b - b_0)$. If $|b_0| < \dfrac{2}{3\sqrt{3}}$ then the line $b = b_0$ meets the slow manifold in three points, two sinks and one source. After the action, the system would be caught by the other sink, violating Quality 3. Therefore this would not be a solution to the problem. This explains why it was necessary to pass from Example 2 to Example 3 in our development.

1.8. APPLICATION TO THE HEARTBEAT

Having obtained the solution in the case of the jump return, we must now test Example 6 as a model for the heartbeat. That is to say, we must identify the variables x, b with measurable qualities and use the differential equation for prediction. Obviously x is going to be the length of the muscle fibre (+ a constant), so that the action represents the contraction, and the jump return represents the relaxation. Meanwhile b is going to be some form of electro-chemical control, and there may be several ways of measuring b. The pacemaker wave will change the control from b_0 to b_1, say, thus triggering off the heartbeat cycle as shown by the dotted line in Figure 10.

We shall go into this in more detail in section 2.1, because the tension in the fibre caused by the blood pressure turns out to be a second important control variable. Meanwhile we observe that one of the simplest ways of measuring b may be to measure the potential across the membrane of the muscle fibre (*see*

23

Heartbeat and nerve impulse

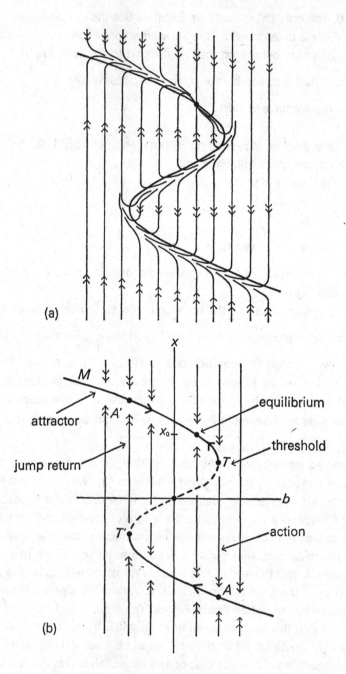

(a)

(b)

FIGURE 9

24

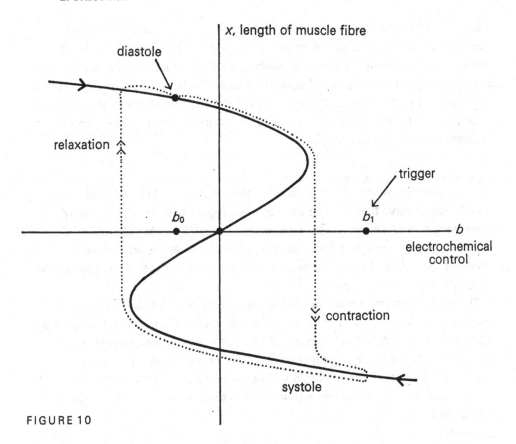

FIGURE 10

section 2.2 and Figure 18). This is not to say that the membrane potential is necessarily a fundamentally important part of the muscle contraction, but only that it may be a convenient artifact for measuring the chemical control, in the same way that colour is a convenient artifact used by a blacksmith to measure the temperature of an iron bar. Alternatively it may be better to measure b by a concentration or a flow of some particular chemical. Indeed we know that muscle chemistry is very complicated, involving not only membrane depolarization, but also changes in permeability, transport of ions, synthesis of actomyosin, breakdown of ATP (adenosine triphosphate) to release the energy for the contraction, amongst other things. Even if we knew the whole story, and were able to represent it by a vast dynamical system in R^n, yet the heartbeat cycle would still be represented by a 1-dimensional path in R^n, consisting of 2 pieces of slow orbit and 2 pieces of fast orbit. And if we chose convenient coordinates x, b such that the projection of the path into the (x, b)-plane was an embedding, then these two coordinates would be sufficient for experimental predictions about the cycle. In fact if our objective is to study the dynamics of the heartbeat rather than the underlying chemistry (which nature may have made unnecessarily complicated), then it may be more efficient to use the R^2-system based on the dynamics, rather than some R^n-system based on approximations to the chemistry.

This completes the discussion on the jump return 3a, and we now turn to the problem of the smooth return 3b.

1.9. DYNAMICAL SYSTEMS ON R^3

We have seen in Lemma 3 that a smooth return is impossible in R^2, and so we try R^3. First observe there can be only one fast eigenvalue. For otherwise, if there were two fast eigenvalues, then the fast foliation would be a family of planes, while the slow manifold would be a curve, as before, and when we consider the points of tangency we should run up against the same argument as before, implying a jump return.

Therefore, for a smooth return there must be only one fast eigenvalue which we choose to be $-1/\varepsilon$, as before, and two slow eigenvalues, which may be real or complex. In the simplest solution it will turn out that they are complex. Choose coordinates x, a, b for R^3. The fast foliation will be the family of lines parallel to the x-axis, while the slow manifold will be a surface cutting the fast foliation transversally near the equilibrium point. First a trivial linear example, analogous to Example 1.

▶ *Example* 7

$$\varepsilon \dot{x} = -x$$
$$\dot{a} = -a + b$$
$$\dot{b} = -a - b$$

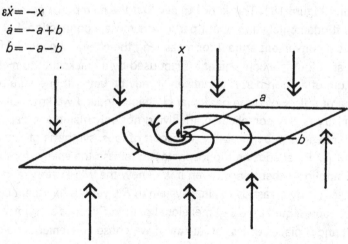

FIGURE 11

26

E. C. Zeeman

The single fast equation determines the slow manifold, which is the plane $x=0$. The pair of slow equations determines that flow on the slow manifold, which is spiralling in towards the origin, because the eigenvalues are $-1 \pm i$. Putting $z=a+ib$, the equations can be written $\dot{z}=-(1+i)z$, with solution $z=z_0 e^{-(1+i)t}$. It is the spiral quality associated with complex eigenvalues that we shall find useful in negotiating the smooth return.

To find the solution in the case of the jump return we laboriously argued our way from Example 1 to Example 6. Now in the case of the smooth return we can telescope these arguments and proceed straight from Example 7 to finding the solution in Example 8.

Let M denote the slow manifold, which will be a smooth surface in R^3. Choose a constant $k<0$, and let C be the plane $x=k$. Then C is perpendicular to the fast foliation. We can use (a, b) as coordinates for C. The letter C stands for *control space*, because sometimes it is convenient to think of a, b as parameters or controls. Let $\chi : M \rightarrow C$ denote the orthogonal projection of M onto C. The only reasons for choosing $k<0$ is that in diagrams it is convenient to draw C as a horizontal plane below M, and not meeting M.

As before, a *threshold T* is a point of tangency of M and the fast foliation, or in other words a point on a fold curve of χ. Therefore Quality 2 implies the existence of a fold curve separating a piece of slow attractor A' from a piece of slow repellor B, as shown in Figure 12. Meanwhile Quality 3 implies the existence of another piece of slow attractor A to catch the action.

Quality 3b implies that A, A' can be connected by an arc in M not crossing any fold curves, in order to achieve the smooth return. The problem is to find the simplest M with these properties. Our criterion of simplicity is to find a fast equation of the form

$$\varepsilon \dot{x} = -f(x, a, b)$$

where f is a polynomial of lowest degree. Therefore M will be given by the equation $f=0$. Now f must be of degree at least 3 in x, because in Figure 12 there are lines parallel to the x-axis meeting M in 3 points. Further f must be of degree at least 1 in a, b because otherwise, if f did not contain either a or b, then we should be back to the 2-dimensional problem, which is impossible by Lemma 3. These lower bounds for the degree of f turn out to be not only necessary, but sufficient, for consider the surface given by the equation

$$f \equiv x^3 + ax + b = 0 \quad .$$

27

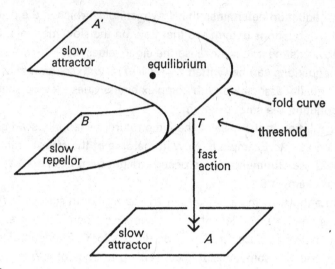

FIGURE 12

This has the required properties, as can be seen from Figure 13. The fold curves are given by :

$$\frac{\partial f}{\partial x} = 3x^2 + a = 0 \quad .$$

When the fold curves are projected onto C they form a cusp, which is obtained by eliminating x from the two equations above, giving

$$4a^3 + 27b^2 = 0 \quad .$$

Outside the cusp M is single sheeted, and inside the cusp M is 3-sheeted. The singularity of the map $\chi : M \rightarrow C$ at the origin 0 is the well-known Whitney *cusp-singularity*. Sometimes we abuse language and call the singularity itself a *cusp*, although in fact M is smooth at 0, and the cusp really exists only in C. In addition to having the Whitney singularity the surface M has the property that the upper and lower sheets are attractors, while the middle sheet is a repellor.

Let us pause for a moment to consider what we are doing. We have found a surface of lowest degree possessing all the required properties. But is this really the 'simplest' example, and is there any virtue in having found the simplest ? The topologist regards polynomials as rather special, and tends to turn his nose up at so crude a criterion of simplicity as choosing a polynomial of lowest degree. Moreover in biology of all subjects we should least expect Nature to be so obliging as to use polynomial equations. So perhaps we ought to consider *all possible* surfaces. Now comes the truly astonishing fact : when we do consider

28

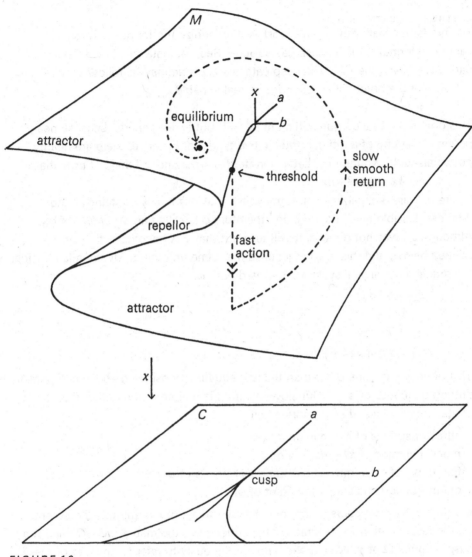

FIGURE 13

all surfaces, not only is this particular surface the *simplest* example, but in a certain sense it is the *most complicated* example, in other words it is the *unique* example. Herein lies the punch of the deep and beautiful catastrophe theory created by René Thom. We must digress to explain precisely the sense in which this surface is unique.

29

Heartbeat and nerve impulse

1.10. THE CUSP CATASTROPHE

The reader familiar with Thom's work will recognize the formulae above as canonical formulae of the cusp catastrophe. See, for instance, David Fowler's paper in this volume [7]. The cusp catastrophe is defined as follows.

Consider the potential function in the real variable x,

$$\varphi_0 = \tfrac{1}{4}x^4 \quad .$$

This potential has a unique minimum at $x=0$ but is not generic, because nearby potentials can be of a different qualitative type. For instance, there are nearby potentials with two minima. Let φ denote the 2-parameter family of potentials

$$\varphi = \tfrac{1}{4}x^4 + \tfrac{1}{2}ax^2 + bx \quad ,$$

where a, b are real parameters. It transpires that this family contains all the different possible types near φ_0, and therefore φ is called an *unfolding* of φ_0. Although φ_0 was not generic, it will turn out that φ is generic, in the sense defined below, and this is what justifies the name unfolding. Meanwhile consider the gradient dynamical system associated with φ,

$$\dot{x} = -\operatorname{grad} \varphi$$

$$= -\frac{\partial \varphi}{\partial x}$$

$$= -(x^3 + ax + b) \quad .$$

This of course is none other than the fast equation considered above in Figure 13, putting the constant $\varepsilon = 1$. Our slow manifold is therefore none other than the set of stationary values of φ. In more detail

attractor surface of M = minima of φ
repellor surface of M = maxima of φ
fold curve of M-origin = cubic stationary points of φ
origin = the quartic stationary point of φ.

We define the *cusp catastrophe* to comprise the set of five things : the potential φ ; the fast equation $\varepsilon \dot{x} = -\operatorname{grad} \varphi$; the surface M ; the map $\chi : M \rightarrow C$; and the cusp-singularity of χ. We use the term ambiguously to refer to any one of these things, provided the context is clear. In a moment we shall define equivalence between potentials, and in some contexts it is convenient to use the term cusp catastrophe ambiguously to refer not only to φ, but also to the equivalence class containing φ, or any member of that class ; in such contexts we refer to φ as the *canonical* cusp catastrophe.

The reason that Thom chooses to use the word *catastrophe* is that in many applications the fast jump between the two attractor surfaces represents a discontinuity in space—time, frequently as a shock-wave (in space) or as a catastrophic change of behaviour (in time).

Consider now the space Ψ of all smooth 2-parameter families of potentials. Since we have fixed x, a, b as coordinates for R^3, any function on R^3 determines an element of Ψ. Therefore Ψ is the same as the space of smooth functions on R^3 (here smooth means that the second partial derivatives are continuous). Given $\psi \in \Psi$, the set M_ψ of stationary values of ψ is defined by $\partial \psi / \partial x = 0$. Given ψ, $\psi' \in \Psi$ define them to be *equivalent* if there exist diffeomorphisms $f : R^3 \to R^3$ and $g : C \to C$ commuting with the projection $R^3 \to C$, and such that $fM_\psi = M_{\psi'}$. Define ψ to be *generic* (or *stable*) if it has a neighbourhood of equivalents in Ψ. It is a theorem that the generic elements form an open-dense subspace of Ψ, and so from the point of view of applied mathematics the non-generic elements can be ignored. It is a theorem that if ψ is generic, then M_ψ is a smooth surface, and any nearby element will have a qualitatively similar surface. For example, the cusp catastrophe φ is generic and therefore any nearby ψ will have a surface M_ψ with the same qualities as M; namely two-fold curves over a cusp, single sheeted outside the cusp and 3-sheeted inside with two attractors and a repellor in the middle. Intuitively any such M_ψ can be obtained by bending M, so as not to disturb these qualities. We can now state the classification theorem.

▶ *Theorem* (*Thom*)

If ψ is generic, then the only singularities of the projection $M_\psi \to C$ are folds and cusps. Locally near any cusp, ψ is equivalent to the cusp catastrophe φ. In other words φ represents the most complicated thing that can happen locally. Of course globally M_ψ may have many cusps and be many sheeted. Also if we enlarge to 3 or more parameters the situation can become more complicated [*see* 1, 2], but for this paper it is sufficient to stick to 2-parameters. Therefore the theorem is the key mathematical fact behind our whole approach. This ends the digression on the cusp catastrophe.

We now go back to our main problem of finding the simplest dynamical system for the smooth return. If we consider *all possible* generic fast equations, it follows from Thom's theorem that the slow manifold must be a smooth surface with only folds and cusps. It would be possible to embed Figure 12 in a surface without cusps (for consider a looped hosepipe) but this would be both topologically more complicated, and biologically more complicated, in the sense of section 1.4 as a development from Example 7. Therefore the criteria of topological and biological simplicity imply the existence of a cusp. We have seen in Figure 13 that one cusp is sufficient, and by Thom's theorem any cusp is equivalent to the canonical cusp catastrophe. Therefore we are justified both mathematically and biologically in choosing the canonical fast equation

$$\varepsilon\dot{x} = -(x^3 + ax + b) \quad .$$

Finally we come to the slow equations. Normally the cusp catastrophe has no slow equations, because a, b are regarded merely as parameters or controls. However when we add slow equations this changes the nature of the variables a, b by giving them a dynamic role, and if x appears in the slow equations this can be regarded as a form of feedback on the cusp catastrophe.

There is nothing unique about the slow equations. In section 2.7 we shall take pains to fit the Hodgkin–Huxley data, and shall derive relatively complicated slow equations, one being an electrical capacitance equation, and the other an *ad hoc* piece of data fitting. But our present objective is to make them as simple as possible in order to provide conceptual insight. Therefore we chose the slow equations to be linear with complex eigenvalues, and we fix the coefficients so as to spiral the slow return smoothly round the cusp.

▶ *Theorem 2*

There exists a dynamical system on R^3 possessing Qualities 1, 2, and 3b. The simplest example is given in Example 8.

▶ *Example 8*

$$\varepsilon\dot{x} = -(x^3 + ax + b)$$
$$\dot{a} = -2a - 2x$$
$$\dot{b} = -a - 1$$

The equilibrium is given by putting $\dot{x} = \dot{a} = \dot{b} = 0$, and is the point $x = 1$, $a = -1$, $b = 0$. To first order in ε, the eigenvalues at the equilibrium point are $-2/\varepsilon$ and the complex cube roots of 1.

Assuming that the trigger increases b, then the threshold is reached when $b = 2/3\sqrt{3}$. The smooth return is shown in Figure 14. The figure shows the flow on the slow manifold as seen from above. The words 'as seen from above' mean that inside the cusp the top attractor surface hides both the middle repellor and the bottom attractor surface. This explains why only the right side of the cusp is visible, and why the flow appears to be discontinuous across this fold line because the left side refers to the upper surface and the right side to the lower. In fact the flow is smooth everywhere.

1.11. APPLICATION TO THE NERVE IMPULSE

If we are to apply Example 8, or a refinement of it, to the nerve impulse, we must identify the dynamical variables x, a, b with measurable quantities. We have

FIGURE 14

the same proviso as in section 1.8, that the experimentally most convenient qualities to measure may in fact be only artifacts, mirroring the dynamics, and we return to this point more deeply in section 2.10.

The natural candidate for b is the membrane potential, because we have assumed that the trigger is given by changing b, and what triggers off the nerve impulse is a depolarization of the membrane. The two other most important variables have been pinpointed by Hodgkin–Huxley [4] as the permeabilities of the membrane to sodium and potassium ions. Of all the changes that occur during the nerve impulse, the most dramatic is the sharp rise in sodium conductance at the beginning of the impulse (*see* Figure 27), and so it is natural to identify this with the action, and to correlate sodium conductance with $-x$. This leaves a to be identified with potassium conductance, and sure enough the latter does not begin to change until after the action has finished, and then it rises and falls smoothly during the smooth return (*see* Figure 27). Therefore Figures 12 and 13 present all the correct qualitative features observed during experiments. However Example 8 is too elementary to predict the correct quantitative features, and in order to build a better quantitative model (diffeomorphic to Example 8), we need to go into the physiology in more detail in section 2.4. This completes the first half of the paper and our objective of deriving qualitative mathematical models of the three dynamic qualities. For the second half of the paper we return to the physiology of the heart.

Part Two

2.1. TENSION AND BLOOD PRESSURE

The purpose of this section is to derive the *local heartbeat equations* from Example 6, allowing for the tension in the muscle fibres due to blood pressure, and to explain a variety of associated phenomena including Starling's Law.

In 1957 Boris Rybak [8] originated the following experiment. I am indebted to him for demonstrating it to me. If the heart (of a frog say) is taken out then it stops beating. However, if it is then cut open into a flat membrane and stretched under slight tension, it then starts to beat again and will continue for some hours. If the tension is relaxed the beating stops. Alternatively if the pacemaker is removed then the beating stops. (This is not quite true because there is a weak secondary pacemaker, and if both pacemakers fail then after a while spontaneous new ones may arise, which are called ectopic pacemakers.) Therefore both factors are necessary for the heartbeat, tension and the pacemaker wave. The tension *in situ* is of course provided by the blood pressure. However the pressure is different in different parts of the heart, and varies during the beat cycle, and so we must examine the global structure.

The heart has four chambers and four valves. At first sight this seems unnecessarily complicated, because a pump needs only one chamber and one valve, but if we examine the pressures involved the reasons become clear. There are two circuits for the blood, one round the lungs to pick up oxygen, and the other round the body to deliver it. The first is a low pressure circuit, so as not to damage the delicate membrane in the lungs, while the second is a high pressure circuit, in order to get down to the feet, through the capillaries, and up again. The right side of the heart is the low-pressure pump for the lungs, and the left side is the high-pressure pump for the body. The maximum/minimum pressures that occur in each part of the heart during normal heartbeat are shown in Figure 15 (in millimetres of mercury). We follow the usual medical convention of interchanging left and right in the picture, because we imagine we are looking at someone else's heart.

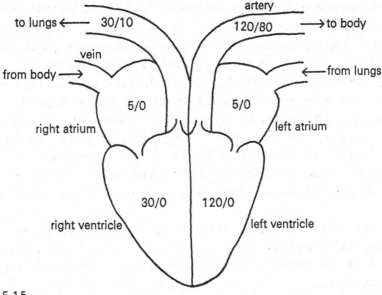

FIGURE 15

Each pump has a main pumping chamber called the ventricle, with an inlet and an outlet valve. It has to have an inlet valve to prevent flowback up the veins while pumping, and an outlet valve to prevent flowback from the arteries while filling. Now comes what would be a problem if there were only one chamber : since the heart is made of non-rigid material it only has the power to push out, and no power to suck in (unlike the lungs which have the ribs). Therefore it would have to depend for filling upon the very feeble venous return pressure. To get a good pump it is necessary to completely fill the ventricle (see Starling's Law below), and so Nature has installed a small prechamber called the atrium, whose job it is to pump gently beforehand, just enough to fill the ventricle, but not too much to cause a flowback up the veins. In fact 60 per cent of the filling of the ventricle is done by the feeble venous pressure before the atrium begins to beat (see [9, p. 107] and the last diagram in Figure 20), and so the atrium does only the final topping up, and consequently does not need to be as big as the ventricle.

The pacemaker sits at the top of the atria, and the pacemaker wave first spreads slowly over the atria, causing them to contract and push the blood into the ventricles. As the inlet valve closes, the wave focuses at a node in between the atria and ventricles, and then spreads rapidly over each ventricle carried by fast fibres (the bundle of His) causing the whole ventricle to contract simultaneously and deliver a big pump of blood down the arteries. Then the muscles relax in the same order, and the cycle repeats itself. The way the left ventricle is made into a high-pressure pump is to have a thicker criss-cross structure of muscle fibres.

Another interesting feature, known as *Starling's Law of the Heart* [10, p. 122], is that the more the muscle fibres are stretched before beating, the more forcible is the beat. It is an excellently designed system for meeting emergencies : suppose fear or rage causes adrenalin to be injected into the blood stream ; then the adrenalin causes the arteries to contract and the pulse rate to increase, which in turn causes the blood pressure to rise, and the atria to push more blood into the ventricles ; finally Starling's Law describes how the stretched ventricles then give a bigger beat, overcoming the increased arterial back-pressure and circulating the blood faster.

On the other hand if the ventricles are overstretched beyond a certain point, as can happen for instance when a person with high blood pressure receives a sudden shock, then the heart may fail to beat, or only beat feebly, and cardiac failure may ensue.

At the other end of the scale, if the bloodstream is made to bypass the heart during an operation, so that there is no longer any blood pressure in the heart, then the beat becomes sluggish, and the heart tends more to heave without altering much in size. This phenomenon is similar to the Rybak experiment, in which the beat ceases when the tension drops.

E. C. Zeeman

Summarizing, the individual muscle fibres can behave very differently, depending upon circumstances, and where they are situated. We propose that most of
this difference in behaviour can be attributed to difference in tension, caused by
varying blood pressure, and we shall show that by a slight addition to the
equations of Example 6 we can explain a large variety of phenomena, including :
(1) Rybak's experiment, and the sluggish beat of the bypassed heart ;
(2) the small atrial beat, followed by the squeezing action of blood through the
inlet valves ;
(3) the large ventricular beat, obeying Starling's Law ;
(4) overstretching causing cardiac failure.

We argue as follows. Rybak's experiment shows that the heart can stop beating
even though the pacemaker is still going. Therefore if a fibre is no longer under
tension the pacemaker wave no longer triggers a sudden contradiction. In other
words if the tension drops, the threshold must disappear. Conversely as the
tension rises the threshold appears. Now we originally constructed the threshold
in Example 5 by changing the sign of x in the fast equation. Therefore the natural
place to put the tension is as the coefficient of x in the fast equation. Consequently we are led, once again, to the cusp catastrophe :
$$\dot{x} = -(x^3 + ax + b)$$
where this time (up to multiplicative and additive constants) x=length of fibre,
b=chemical control, $-a$=tension.

From the point of view of catastrophe theory a, b are the two parameters of
control in the control space C. In Figure 16 the slow manifold M illustrates how
these two controls determine the length of the muscle fibre. Without such a
picture it is difficult to visualize how the two controls interrelate (as students of
anatomy will confirm). The feedback on M is determined by the slow equation,
to which we now turn our attention.

Suppose that the chemical control b takes values b_0 in diastole and b_1 in
systole. More precisely $b=b_0$ is the equilibrium position, and the trigger moves
b from b_0 to b_1, after which the slow equation returns b to b_0. Therefore during
diastole b is indeed fixed at b_0, but during systole b is changing, with a maximum
at b_1, as illustrated in Figure 10.

We shall assume that $b_0 < 0 < b_1$. The justification for this assumption is that it
leads to the simplest explanation for the variety of phenomena, but of course it is
a hypothesis that needs to be tested experimentally. It may be that b_0, b_1 depend
upon other factors (for example, oxygen lack may increase b_0), and an investigation into the consequences of changing these constants might give insight into

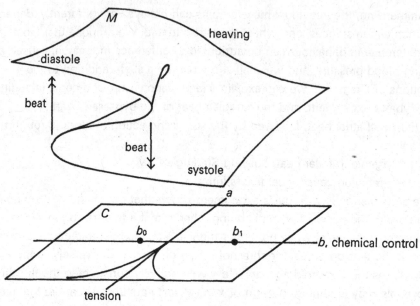

FIGURE 16

pathological behaviour of the heart. But for now let us assume $b_0 < 0 < b_1$. Under given tension a, the diastole length x_a of the fibre will be given by the equilibrium condition $\dot{x} = 0$. Therefore x_a is the solution of

$$x_a{}^3 + ax_a + b_0 = 0 \quad ,$$

lying on the upper diastole surface, in other words satisfying the inequality

$$x_a > \sqrt{-\frac{a}{3}} \quad .$$

We can now write down the slow equation

$$b = x - x_a \quad .$$

We do not postulate a slow equation for a, because the main reasons for tension in a muscle fibre are not the internal chemistry but external forces caused by the blood pressure and the pushing and pulling of the different organs of the heart against each other.

Summarizing : *The local heartbeat equations*

$$\varepsilon\dot{x} = -(x^3 + ax + b)$$
$$b = x - x_a$$

where (up to multiplicative and additive constants) $x =$ length of muscle fibre, $b =$ chemical control (possibly membrane potential), $-a =$ tension.

The behaviour of a fibre under a particular tension a_1 is given by taking the

38

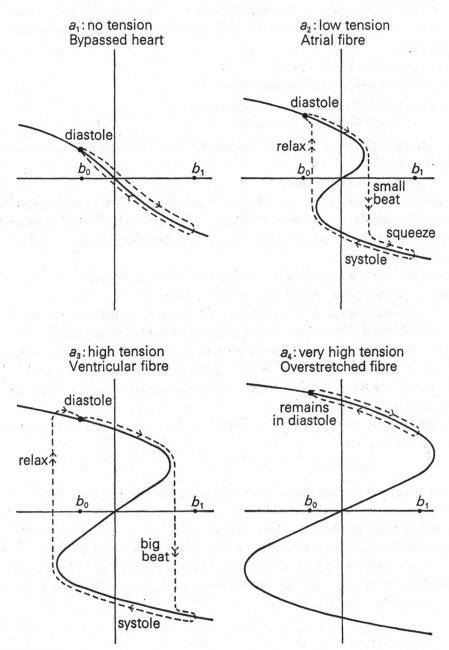

a_1: no tension
Bypassed heart

diastole

b_0 b_1

a_2: low tension
Atrial fibre

diastole

relax

b_0 b_1

small
beat

squeeze

systole

a_3: high tension
Ventricular fibre

diastole

relax

b_0 b_1

big
beat

systole

a_4: very high tension
Overstretched fibre

remains
in diastole

b_0 b_1

FIGURE 17

39

Heartbeat and nerve impulse

section $a=a_1$. To explain the variety of phenomena listed above, we draw four sections in Figure 17 for four values $a_1 > a_2 > a_3 > a_4$ corresponding to the four situations :

a_1 : no tension, bypassed heart ;

a_2 : low tension, beat of an atrial fibre ;

a_3 : high tension, beat of a ventricular fibre ;

a_4 : very high tension, cardiac failure.

In each case the slow manifold is indicated by a thick line and the beat cycle by a dotted line.

Notice the following features :

Case (1). Each fibre does contract a little, but slowly rather than sharply. Therefore the pacemaker wave induces a muscular reaction that can be observed as a wave of sluggish contraction spreading over the heart, rather than a sharp contraction.

Case (2). After the small beat the atria continue to contract slowly for a while, thereby squeezing the blood through the inlet valve.

Case (3). After relaxation the ventricles expand beyond equilibrium, helping them to do most of the filling before the next beat begins. If the tension is increased, then the S-shape of the slow manifold becomes more pronounced, so that the fibre is stretched longer in diastole and contracted shorter in systole. Hence the action is bigger and faster, making the beat more forcible, and thereby explaining Starling's Law.

Case (4). This case occurs if the tension is sufficiently large. Mathematically the condition on a_4 is :

$$a_4 < -(27b_1^2/4)^{\frac{1}{3}} .$$

The condition moves the threshold beyond b_1, so that the trigger no longer reaches the threshold ; consequently the fibre remains in diastole. If a few of the ventricular fibres manage to contract first, then they help to overstretch the rest and prevent the rest from contracting, resulting in a feeble ventricular beat. Meanwhile the atria increase the strength of their beat by Starling's Law, further overfilling the ventricles, and possibly causing permanent damage to the latter by overstretching.

2.2. THE PACEMAKER WAVE
The mechanism underlying the pacemaker wave P is not fully understood, even though it is the basis of the extensive field of electrocardiography. We propose a

tentative model for P, and argue the importance of separating the mathematics of P from the local heartbeat equations.

P is primarily an electrical phenomenon, probably because electrons are the only physical objects that can traverse tissue at sufficient speed. When P reaches an individual muscle fibre it causes a depolarization of the cell membrane potential, which triggers off the sequence of chemical events leading to the contraction. Figure 18 is taken from Sampson Wright's *Applied Physiology* [10, p. 96]. The upper curve is the time-graph of the membrane potential of a frog ventricular fibre, beating at the rate of 1 per second. If we identify the membrane potential with our chemical control, b, then the thin steep rising line of depolarization corresponds to the trigger, changing b from b_0 to b_1 and causing contraction, while the thick descending curve of repolarization corresponds to our slow equation, ending in relaxation.

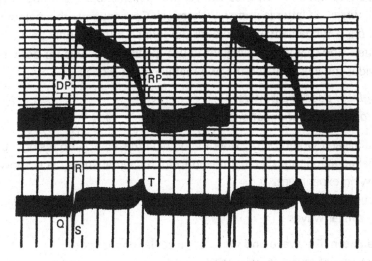

FIGURE 18
A normal cardiac transmembrane potential and its relation to the surface electrocardiogram. D P = depolarization. R P = repolarization. Time lines, 0·1 sec. (*after* Hecht, 1957)

Meanwhile the lower curve is the skin E C G (electrocardiograph), which is the summation over all fibres in the heart. An E C G can be recorded between any pair of points on heart or skin, because the electrical changes are conducted throughout the body fluids, although of course different pairs of points show slightly different readings. The physician uses the E C G to check whether, and when, the atria and ventricles are beating. However we are interested not in the

41

global summation, but in the local membrane potential b, which we can regard as a function of time and position on the heart H. We shall represent the pacemaker wave P as a wave in b.

Now the speed of P depends upon the conductivity of the heart tissue, which is not the same in all directions. Perhaps the best way to model this is to represent H by a surface (with singularities) and represent conductivity by a Riemannian structure on H. We can then define P to be the solitary wave, of amplitude $b_1 - b_0$, emanating from the pacemaker, and propagating over H according to Huyghens' principle, at speed determined by the Riemannian structure. Consequently *the wave front is given by the geodesic flow on H*, emanating at a given moment from the pacemaker.

We summarize the difference between the pacemaker wave P and the muscle fibre contraction. The former is global and the latter local. The former triggers the latter. Both depend upon electrochemical mechanisms, but the former is primarily electrical, while the latter is primarily chemical. Consequently the membrane potential b is a fundamental variable for the former, but may only be an artifact in the latter. However it is convenient to use the same variable for both, because this provides the link between the two mathematical models :

(1) *the global geodesic flow for P* ;
(2) *the local heartbeat equations.*

Mathematically it would be possible to combine these two into a single partial differential equation, but this might lead to fundamental errors, as is shown by the following experiment.

From such a *combined* equation we should be able to extract two mathematically similar waves :

P : the pacemaker wave and wave of muscular contraction :
Q : the wave of muscular relaxation.

In the normal heartbeat of man Q follows P after about 1/5 second, and in the frog example in Figure 18 the interval is about 1/2 second. If we make a cut across the heart tissue just before the arrival of P, then this would stop P, because P propagates by Huygens' principle, and so could not jump across the cut. On the other hand if we were to make the cut just before the arrival of Q, then this would *not* stop Q, because Q propagates not by Huyghen's principle, but as a wave of local events occurring 1/5 second after P. The mathematical similarity of the two waves would deceptively obscure their physical difference.

2.3. CONCLUSION AND EXPERIMENTS

There are three types of complexity in the heart :

(1) the global complexity of the anatomy ;

(2) the local complexity of the underlying biochemistry ;

(3) the dynamic complexity of the beat.

Most research has been directed at unravelling (1) and (2), with the ultimate objective of understanding (3). What we have tried to do is bypass (1) and (2) and isolate a central simplicity in the mechanical behaviour of muscle fibres. We have expressed this simplicity mathematically in the local heartbeat equations, and have shown that a large part of the dynamic complexity can be deduced fairly simply from these equations, and does not depend upon the complexity of (1) and (2). This does not mean we escape from (1) and (2) ; on the contrary the next thing to do is to relate (1) and (2) to the central simplicity.

First we must test the equation against the underlying chemistry. For example, testing fibre length against membrane potential and tension should verify the qualitative features of the cusp catastrophe, and provide the real refined version of the fast equation, or a functional relationship between the membrane potential and chemical control. Testing the fibres dynamically would give a refined version of the slow equation. It may transpire that the slow equation is determined by ionic flow, as in the nerve impulse, in which case it could be misleading to have fibre length in the slow equation. But the point is that the equations provide a nucleus of theory around which to design a series of experiments.

Secondly we must relate the local simplicity of fibre behaviour to the global geometry of the anatomy. We do this by visualizing how the heart moves over the surface of the cusp catastrophe during the beat cycle, and in this way we shall derive the beat from a smooth motion in the control space (*see* Figure 20). The objective here is to gain a better conceptual understanding of the heartbeat as a whole, which should lead not only to design of experiments but also to a better understanding of heart failure, various forms of heart blocks, and perhaps improved diagnosis.

In Figure 17 we have illustrated the difference between atrial and ventricular fibres. The picture of the atrial fibre is reasonably accurate, because the blood pressure in the atria does not vary greatly, and therefore neither does the tension in atrial fibres. However the picture of the ventricular fibre is an oversimplification because the ventricular pressure varies considerably during the beat cycle [10, p. 107].

At a given time t, each fibre is subject to some tension and chemical control,

43

Heartbeat and nerve impulse

and so there is a map $f_t : H{\to}C$ from the heart H to the control space C. The muscular state of H is given by lifting f_t to a map $g_t : H{\to}M$, where M is the cusp catastrophe surface, or slow manifold. Lifting means that Figure 19 is commutative. Although χ^{-1} is 3-valued over the interior of the cusp, the lifting is uniquely defined by past history and the delay convention. The *delay convention* (which is an immediate consequence of the fast equation) says to each fibre 'keep on the upper or lower surface as long as possible, and only jump onto the other one if necessary, because of crossing a threshold'. In Figure 20 we illustrate a lifting, representing symbolically one of the ventricles by a triangle, the corresponding atrium by a circle, and the node between them by a line.

The diagram shows that even though f_t is continuous, g_t may not be. The pacemaker wave has already carried the top half of the atrium past the threshold,

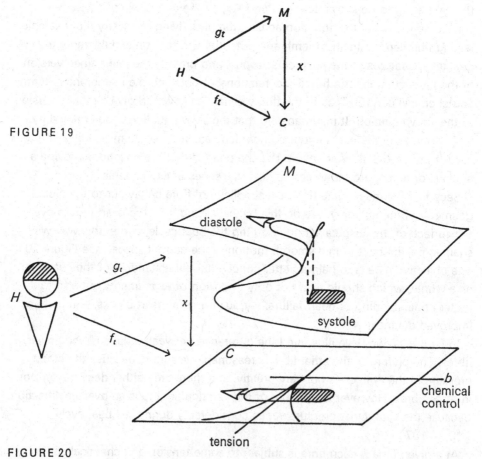

FIGURE 19

FIGURE 20

44

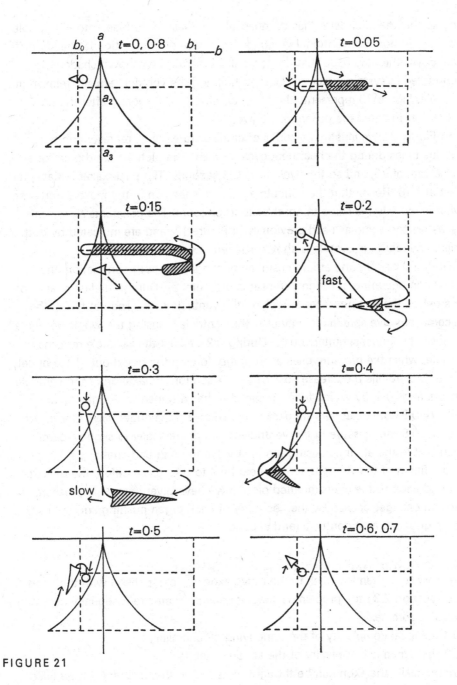

FIGURE 21

and so the image under g_t has dropped off the diastole surface onto the systole surface. In more conventional language, the wave of contraction is already half way down the atrium, so that the top half of the atrium (shown shaded) is contracted in systole, while the bottom half and the ventricle are still relaxed in diastole. Figure 20 represents the situation about 1/20 second after the pacemaker has initiated the pacemaker wave.

In Figure 21 we illustrate a series of eight pictures of C, representing f_t at various times during the heartbeat cycle in man. The pictures are drawn for a pulse-rate of 75, and so the cycle lasts 0·8 seconds. The pacemaker initiates the beat at $t=0$. Recall that the variable a measures tension in the muscle fibres, or what is roughly the same thing, the blood pressure. The constants b_0, b_1 and a_2, a_3 are the same as used previously in Figure 17, and are indicated by dotted lines. The region in systole is shown shaded in each case.

During $0 < t < 0·2$ the atrium is slowly beating, with both contraction and relaxation occurring at roughly the same pressure, a_2. During the last quarter of this period the pressure rises sharply in the ventricle to a_3. Then at $t=0·2$ the pacemaker wave spreads rapidly over the ventricle, causing the whole ventricle to contract almost simultaneously. During $0·2 < t < 0·4$ the ventricle remains in systole, while the pressure rises even higher forcing the blood out of the outlet valve. Meanwhile b decreases by the slow equation. Relaxation of the ventricle occurs at $t=0·4$, by which time the pressure has dropped to a_3 again. During $0·5 < t < 0·8$ both chambers are back in diastole with the outlet valve closed. During $0·6 < t < 0·8$ the pressure in the ventricle drops dramatically to slightly lower than that in the atrium, causing the inlet valve to open and enabling 60 per cent of the filling of the ventricle for the next beat to take place before the beat starts.

In addition to the oversimplified picture we have given, there are pressure and tension changes caused by the geometry of each organ pushing and pulling on its neighbours, as it contracts and expands.

2.4. PROPAGATION AND TIMING OF THE NERVE IMPULSE

We now turn attention back to the nerve axon. By comparison with the heart (*see* section 2.3) the problem is simpler because there are essentially only two types of complexity :

(1) the local complexity of the underlying biochemistry ;

(2) the dynamic complexity of the action potential.

Anatomically the axon can be thought of as a long thin cylindrical tube filled with a fluid called axoplasm. To get an idea of size, a typical axon in man might be 1 μ in diameter (1 $\mu=10^{-3}$ millimetres) and a few millimetres long. In different organs and in different species axons can vary in diameter from 0·1 μ to 500 μ, and in length from a fraction of a millimetre to several metres [see 5, p. 15]. For example the data in Figures 3, 22, and 24 were recorded from a giant squid axon of diameter 476 μ.

The axoplasm can conduct electric current, but the wall membrane acts as an insulator. Most axons in man are surrounded by myelin sheathing, which reinforces the insulation, and increases the velocity of the action potential. Every millimetre or so, there are gaps in the myelin sheathing called the nodes of Ranvier, which have different electrochemical properties [see 5, chapter 4], but as a first approximation we shall ignore them, and assume that the walls are homogeneous.

As explained in section 1.11 the important variables are the membrane potential V, and the membrane permeabilities to sodium and potassium ions. Changes in V cause, and are caused by,

(1) flow of electrons along the axon ;
(2) flow of ions through the membrane.

As in the case of the heart (see section 2.2) we shall model these two processes mathematically by

(1) a global propagation wave ;
(2) the local nerve impulse equations.

We have not discussed the propagation wave yet and shall do so in detail in section 2.8 below, and compute the velocity. The local equations we have already discussed qualitatively in section 1.11 and Example 8, but we now get down to a detailed quantitative version in section 2.7 below.

Meanwhile we emphasize the importance of keeping the two equations separate, and not combining them into a single partial differential equation. Otherwise we should be able to extract from the combined equation two mathematically similar waves, about 1 millisecond apart :

P : propagation wave, depolarizing the membrane,

Q : repolarization wave.

The mathematical similarity would obscure the physical difference between P and Q, which is demonstrated by cutting the axon just before the arrival of P, which would stop P, and cutting it just before Q, which would not stop Q. The reason is that Q is a wave of local events, originally triggered by P, but thereafter locally determined by the local equations.

This is verified experimentally by an elegant result of Hodgkin and Huxley [4],

47

as follows. Trigger an impulse in an isolated segment of axon, by means of a short shock administered simultaneously to the entire segment ; then the segment displays the normal action potential without any propagation effects along the axon. In other words it obeys the local equations only. The result is shown in Figure 22, which is taken from Hodgkin [5, p. 65]. The numbers attached to the curves denote the shock strength in $m\mu$ coulomb/cm^2. If the shock is small there is a delay before the action potential is triggered, which we shall explain by means of a saddle point (*see* section 2.7 below, and section 2.10, Remarks 6 and 7).

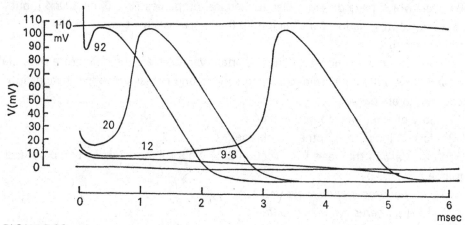

FIGURE 22

Hodgkin [5, p. 14] suggests a very good metaphor in order to stress the fact that the energy for the nerve impulse is released locally rather than propagated globally : he points out that nerve conduction resembles the burning of a fuse of gunpowder rather than the propagation of an electric signal along a cable. In our model, the propagation wave is analogous to the flame running along the fuse setting it alight, while the local equations are analogous to the way it burns locally. There is only one aspect in which this metaphor can be misleading, and that is in the matter of timing. It is worthwhile estimating roughly the time intervals involved.

Suppose we have a segment of axon 1000 times as long as it is wide, for example 1 μ in diameter and 1 mm in length. Suppose that the velocity of propagation is 100 metres/sec. Then the propagation wave traverses this segment in 10^{-2} msec. Meanwhile the complete action potential takes about 1 msec, which is 100 times as long. To put it another way, the wave length is 10 cm,

48

which is much longer than most axons. Of course in different organs and different species these times vary — for example, Figures 3 and 22 refer to a giant squid axon, in which the velocity is only about 20 metres/sec. But the point to be taken is that propagation along the axon is much swifter than the local potential and permeability changes, whereas in the gunpowder fuse it is the exact opposite, because propagation along the fuse is much slower than local burning. Mathematically the timing is very important, because it implies that the propagation wave dominates only at the beginning, after which the local equations dominate. This enables us to keep the two pieces of mathematics fairly separate, and justifies the conceptual device of regarding the end of the propagation wave as the trigger, or initial conditions, for the local slow equations (*see* section 2.9 *below*).

2.5. VOLTAGE CLAMP DATA

The next three sections are devoted to deriving the *local nerve impulse equations*, which are stated in section 2.7. These are a quantitative refinement of the qualitative model Example 8 developed in sections 1.10 and 1.11. We use the electrical theory and voltage clamp data of Hodgkin and Huxley [4, 5].

The voltage clamp technique evolved by Cole consists of isolating a segment of axon, clamping the voltage difference across the membrane to various fixed values by means of electrodes inside and outside, and measuring the currents carried by the resulting flows of potassium and sodium ions across the membrane. From this can be calculated the large changes that occur in the potassium conductance g_K and sodium conductance g_{Na} of the membrane. Meanwhile the chlorine conductance g_{Cl}, and that of other ions, remain relatively small and constant. What happens is illustrated on the cusp catastrophe in Figure 23. We have identified coordinates

$b \sim V$, voltage

$a \sim g_K$, potassium conductance

$-x \sim g_{Na}$, sodium conductance,

where $\sim$ indicates linear or functional relation between.

When the clamp is switched on, the potential V jumps from resting potential 0 to clamp potential V_c, and so the state is displaced from equilibrium E parallel to the V-axis to the position F in the clamp plane $V = V_c$. Then the fast equation carries the state rapidly to the point G on the slow manifold. Finally the slow equations (or more precisely the $\dot{a}$-component of the slow equations) carry the state from G to H, where H is the point where $\dot{a} = 0$. In other words H is the

49

Heartbeat and nerve impulse

point whose slow vector is perpendicular to the clamp plane. Therefore H is the equilibrium position under the clamp, but if the clamp is released then the state returns to equilibrium E along the dotted flow lines of the slow equations. The effect of the voltage clamp on the sodium conductance g_{Na} is a fast increase along the path FG, followed by a slow decrease GH, while the effect on the potassium conductance g_K is a delay FG, followed by a slow increase GH.

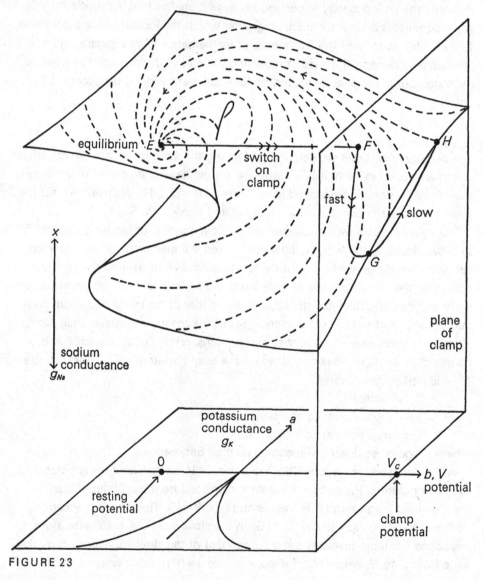

FIGURE 23

This agrees qualitatively with the Hodgkin–Huxley data, shown in Figure 24, which is taken from Hodgkin [5, p. 61], and given in more detail in [4].

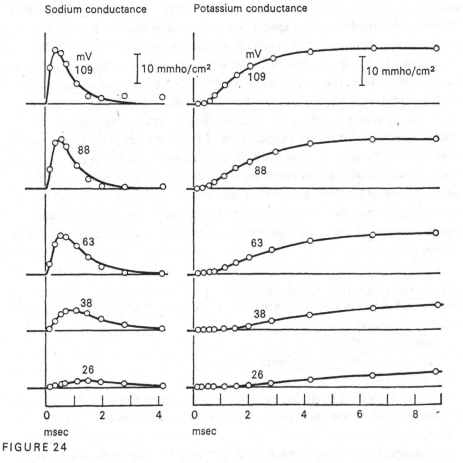

Sodium conductance   Potassium conductance

FIGURE 24

In Figure 24 the numbers indicate the voltage clamp, the circles are the experimental data, and the smooth curves are solutions of Hodgkin and Huxley's own theoretical equations [4, p. 518] designed to fit the data. Our aim is to find alternative equations by fitting the cusp catastrophe to the data, and there are two possible procedures.

Procedure (*1*). Use the data from the time of maximum sodium onwards to plot g_{Na} as a function of V and g_K. Then see if the resulting fragment of surface can be extended smoothly to a surface differentiably equivalent to the cusp catastrophe. The data gives the $\dot{a}$-component of the slow flow on the fragment of surface, while the b-component is given by the electrical capacitance equation (*see* section 2.6 *below*). The advantage of this procedure is that the construction of the fragment is direct, but the disadvantages are the non-uniqueness of the extension, and the unlikelihood of being able to express the results algebraically. Therefore we could not write down equations. However for the purposes of prediction the results could probably be encoded in a computer programme. *Procedure* (*2*). Retain the canonical fast equation, and hence the canonical cusp catastrophe surface. Then fit the data to this canonical surface by juggling the position of the equilibrium point and the functional relations between b, a, x and V, g_K, g_{Na}. Construct an algebraic equation for $\dot{a}$ to fit the data, and use the electrical equation for b. The disadvantage of this procedure is that it may not be possible to fit the data precisely, and the juggling does not give a unique answer. However the advantages are that we are working on the canonical surface, which is justified by Thom's theorem (*see* section 1.10), and we obtain usable algebraic equations. In addition the form of these equations may give insight into the underlying chemistry (*see* section 2.10). Therefore we shall attempt the second procedure.

2.6. ELECTRICAL EQUATION FOR IONIC FLOW

We review briefly the electrical theory of Hodgkin and Huxley [4, 5]. Recall the notation of section 1.1.

v = membrane potential
= potential inside axon − potential outside,
$V = v - v_r = v + 65$

where v_r is the resting potential. Thus the resting potential or equilibrium position is given by $V = 0$. In the computation below we use the value $v_r = -65$ mV [*see* 4, p. 520].

It is observed that the concentration of certain ions is very different in axoplasm and in blood [5, p. 28]. Consider first potassium ions, and let K_i, K_0 denote concentrations inside and outside the axon. The observed ratio is about $K_i/K_0 = 20$. This ratio is maintained by metabolism, and the small flows that occur during the action potential have negligible effect.

Now if the membrane were permeable to potassium, and if the voltage were clamped at $v = 0$, then diffusion would cause an *outward* flow of potassium until the ratio was reduced to 1. However in the resting state the membrane is *not* permeable, and $v_r \neq 0$. Even if the membrane were permeable, the fact that $v_r < 0$ would tend to cause an *inward* flow of potassium ions, since the ions are positively charged. Define v_K to be the theoretical potential at which the outward and inward flows would just balance ; more precisely if the membrane were permeable, and if the voltage were clamped at $v=v_K$, then diffusion would produce an equilibrium at the observed ratio. The value of v_K can be computed by the Nernst formula [5, p. 30] :

$$v_K = \frac{RT}{F} \log \frac{K_0}{K_i} \quad ,$$

where $R =$ gas constant, $T =$ absolute temperature, and $F =$ Faraday. The values we use in the computations below are shown in Table 1 [see 5, p. 28; and 4, p. 520]. The values of b in the last column refer to section 2.7 below.

TABLE 1

Ion		Concentration mmole/kg		v	V	b
		inside	outside			
Potassium	K+	400	20	$v_K = -77$	$V_K = -12$	$-1\cdot4$
Sodium	Na+	50	440	$v_{Na} = +50$	$V_{Na} = 115$	$4\cdot95$
Chlorine	Cl-	90	560	$v_{Cl} = -46$	$V_{Cl} = 19$	$0\cdot15$

The other ions have negligible effect and so we ignore them.

At any given time the flow of potassium ions across the membrane is proportional to

$$v - v_K = V - V_K \quad ,$$

and depends also upon the potassium permeability. The simplest way to measure permeability is electrically, by measuring the *sodium conductance* per unit area, which is defined as follows. If I_K denotes the outward electric current per unit area carried by the outward flow of potassium ions, then define

$$g_K = \frac{I_K}{V - V_K} \quad .$$

Therefore the total outward flow of ionic current per unit area

$$= I_K + I_{Na} + I_{Cl}$$
$$= g_K(V - V_K) + g_{Na}(V - V_{Na}) + g_{Cl}(V - V_{Cl}) \quad .$$

Now the membrane acts as a condenser, of capacity c per unit area, say, and so there is an apparent outward current of $c\dot{V}$, due to storage of charge on the surfaces of the condenser. Therefore if I denotes the total outward flow of current across the membrane per unit area, then

$$I = c\dot{V} + g_K(V - V_K) + g_{Na}(V - V_{Na}) + g_{Cl}(V - V_{Cl}) \quad .$$

53

To balance this outward flow across the membrane there must be an internal flow of current along the axon (carried by electrons moving through the axoplasm). If we make the *local hypothesis* that the flow along the axon is the same at all points of the axon (although it may be varying in time), then no charge accrues at any point, and so we obtain the *local electrical equation* $I=0$. The local electrical equation provides the slow equation for b in the next section.

Note that $I \neq 0$ during the propagation period, as we shall see in section 2.8, but as soon as the latter is over then $I=0$, and the axon behaves in unison according to the slow equations.

2.7. LOCAL NERVE IMPULSE EQUATIONS

We are now in a position to state the local equations. As notation, let $[y]_\pm$ denote the functions

$$[y]_+ = \begin{cases} y, & y \geqslant 0 \\ 0, & y \leqslant 0 \end{cases} \qquad [y]_- = \begin{cases} 0, & y \geqslant 0 \\ y, & y \leqslant 0 \end{cases} .$$

The functional relations between physical variables and canonical coordinates are chosen to be :

Potential	$V=(20b+16)$ mV
Potassium conductivity	$g_K = 2 \cdot 38[a+0 \cdot 5]_+$ mmho/cm².
Sodium conductivity	$g_{Na} = (4[x+0 \cdot 5]_-)^2$ mmho/cm².

We use constants :

Chlorine conductivity	$g_{Cl} = 0 \cdot 15$ mmho/cm² [11, p. 366]
Membrane capacity	$c = 1$ μFarad/cm² [4, p. 520].

Define the *local nerve impulse equations* to be :

$$\dot{x} = -1 \cdot 25(x^3 + ax + b)$$
$$\dot{a} = (x+0 \cdot 06(a+0 \cdot 5))(x-1 \cdot 5a-1 \cdot 67)(0 \cdot 054(b-0 \cdot 8)^2 + 0 \cdot 75)$$
$$\dot{b} = -g_K(b+1 \cdot 4) - g_{Na}(b-4 \cdot 95) - g_{Cl}(b-0 \cdot 15) .$$

The fast equation for $\dot{x}$ is the canonical cusp catastrophe. The slow equation for b is the electrical equation of section 2.6. The slow equation for $\dot{a}$ is an *ad hoc* equation fitting the voltage clamp data : the first factor vanishes at the clamp equilibrium, the second at the resting potential, and the third adjusts the speed to fit that of the data. The *resting potential* or *equilibrium position* occurs when $b = \dot{a} = \dot{x} = 0$, giving

$$b = -0 \cdot 8 \quad \text{and therefore} \quad V = 0$$
$$a = -0 \cdot 4 \qquad\qquad\qquad g_K = 0 \cdot 24 \text{ mmho/cm}^2 \text{ [see 4, p. 509]}$$
$$x = 1 \cdot 07 \qquad\qquad\qquad g_{Na} = 0 .$$

Notice that sodium conductance remains zero until $x < -0.5$. (We discuss this sodium cut off in section 2.10, Remark 5.) The propagation wave displaces the state away from equilibrium to the *trigger point* J (*see* section 2.9 *below*) given by :

$$b = 1.2 (V = 40)$$
$$a = -0.45$$
$$x = -1.2 \quad .$$

En route the state crosses the *threshold curve* near

$$b = 0.1 (V = 18)$$
$$a = -0.4$$
$$x = 0.4$$

where it jumps by the fast equation from the upper surface to the lower surface of the slow manifold, switching on the sodium flow.

FIGURE 25

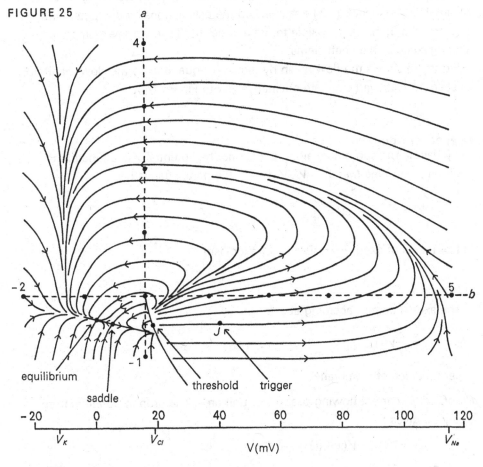

Heartbeat and nerve impulse

There is also a *saddle point* of unstable equilibrium on the slow manifold at

$$b = -0.55$$
$$a = -0.45$$
$$x = 1$$

which explain the phenomena in Figure 22, that perturbations $V \leqslant 8$ are stable and slowly return to equilibrium, but perturbations $V \geqslant 9$ are unstable and slowly increase, until after about 3 msec the threshold is crossed and the normal action potential ensues. There are more remarks about the saddle point below in section 2.10 (6) and (7).

The calculations in this paper based on the equations have been done by hand, and have been used to draw accurate diagrams of the flow in Figure 25, the action potential in Figure 26 and the conductivity changes in Figure 27. Meanwhile Woodcock [12] has checked the calculations by computer, and suggests that it may be possible to get a better fit of the voltage clamp data by altering some of the coefficients.

Figure 25 shows the flow given by the slow equations on the slow manifold, as seen from above (compare qualitatively with Figure 14).

2.8. PROPAGATION WAVE

Let y denote the coordinate along the axon. The potential V is a function of position y and time t, and let V', $\dot{V}$ denote the partial derivatives

$$V' = \frac{\partial V}{\partial y} \ , \qquad \dot{V} = \frac{\partial V}{\partial t} \ .$$

Let $r =$ radius of axon, $\rho =$ resistivity of axoplasm.

▶ *Lemma 1*

The axonal current accruing at y is

$$I = \frac{r}{2\rho} V'' \ ,$$

per unit area of membrane.

Proof. Let $j =$ current flowing past y, per unit area. The drop in potential from y to $y + \delta$ gives :

$$j\rho\delta = V(y) - V(y+\delta) = -\delta V' \ .$$

56

E. C. Zeeman

Therefore

$$j = -\frac{V'}{\rho} \quad .$$

The amount of current flowing into the segment of axon between y and $y + \delta$

$$= \pi r^2 (j(y) - j(y + \delta))$$
$$= -\pi r^2 \delta j'$$
$$= \frac{\pi r^2 \delta}{\rho} V'' \quad .$$

Dividing this by the surface area $2\pi r \delta$ of the membrane of this segment gives the desired result.

▶ *Lemma 2*

If the resting potential is perturbed then the outward ionic current is approximately kV per unit area of membrane, where $k = 0\cdot388$.

Proof. We use the local nerve impulse equations. At equilibrium $\dot{a} = 0$, and if V is perturbed $\dot{a}$ remains small. Therefore we may assume a remains constant at $a = -0\cdot4$. Therefore near equilibrium

$$g_K = 0\cdot238$$
$$g_{Na} = 0$$
$$g_{Cl} = 0\cdot15 \quad .$$

Therefore the ionic flow $= 0\cdot238(V + 12) + 0\cdot15(V - 19)$

$$= 0\cdot388V, \text{ as required.}$$

We assume that the propagation wave starts at $y = 0$ at time $t = 0$, and proceeds with constant velocity θ. Therefore before the wave has reached y, the axon is still in equilibrium :

$$V = 0, \quad y \geqslant \theta t.$$

After the wave has reached y the potential is given by the following theorem.

▶ *Theorem 3*

For $\theta t \geqslant y$, the propagation wave is given by

$$V = A \left\{ \exp\left[\frac{2c\rho\theta}{r}(\theta t - y) \right] - \exp\left[-\frac{k}{c\theta}(\theta t - y) \right] \right\}$$

where A is constant.

Proof. Equating the axonal current of Lemma 1 with the membrane current of section 2.6, and using Lemma 2 :

$$I = \frac{r}{2\rho} V'' = c\dot{V} + kV \quad .$$

Heartbeat and nerve impulse

Since the speed of propagation is constant we can substitute $\ddot{V}=\theta^2 V''$ to obtain

$$\frac{r}{2\rho\theta^2}\ddot{V}-c\dot{V}-kV=0 \quad .$$

This linear equation has solutions of the form $V=Ye^{\lambda t}$, where Y is a function of y, and

$$\lambda=\frac{2c\rho\theta^2}{r} \quad \text{or} \quad -\frac{k}{c}$$

to first order in $1/\theta$, assuming that θ is large. Writing down the general solution, using the fact that it is a wave with velocity θ, and imposing the initial condition $y=0$ when $t=0$, gives the theorem.

We now go back to the observed action potential in Figure 3 (in section 1.1). There is a point J, prior to which there is exponential growth, and after which the slope $\dot{V}$ is constant for a period, and so there cannot be exponential growth. Therefore we assume that *J is the trigger, and prior to J the propagation wave applies.* The values of V, $\dot{V}$ at the critical point J are

$$V_J=40 \text{ mV}$$
$$\dot{V}_J=567 \text{ mV/msec.}$$

From these two readings and Theorem 3 we can deduce the propagation velocity θ, for the particular axon to which Figure 3 refers.

▶ *Corollary to Theorem 3*

$$\theta=21\cdot9 \text{ metres/sec.}$$

Notice that this agrees well with the observed velocity of $21\cdot2$ m/sec (*see* Figure 3), and is slightly better than the Hodgkin–Huxley computation of $18\cdot8$ m/sec.

Proof. After $0\cdot35$ msec the ratio of the growth term to the decay term in Theorem 3 is greater than e^5, and so we can ignore the latter. Therefore

$$\dot{V}=\frac{2c\rho\theta^2}{r}V \quad .$$

Putting in the values at the trigger point J,

$$\theta^2=\frac{\dot{V}_J}{V_J}\frac{r}{2c\rho} \quad .$$

In order to calculate θ in metres/sec we must rewrite the units in terms of metres and seconds, as follows.

$$\frac{\dot{V}_J}{V_J}=14\cdot2/\text{msec}=14\cdot2\times10^3/\text{sec}$$

$$c=1 \ \mu\text{Farad/cm}^2=10^{-2} \text{ Farad/m}^2.$$

58

E. C. Zeeman

For the axon under consideration [see 4, p. 528]

$$\rho = 35 \cdot 4 \text{ ohm.cm} = 35 \cdot 4 \times 10^{-2} \text{ ohm.m}$$
$$r = 238 \mu = 238 \times 10^{-6} \text{ m}.$$

Hence

$$\theta^2 = \frac{14 \cdot 2 \times 10^3 \times 238 \times 10^{-6}}{2 \times 10^{-2} \times 35 \cdot 4 \times 10^{-2}} = 478 \quad .$$

Therefore $\theta = 21 \cdot 9$ metres/sec.

2.9. CALCULATION OF ACTION POTENTIAL

To test the model we now calculate the action potential, that is, the graph of V against time. There are two separate phases : We assume that the action potential is determined

(1) before J by the propagation wave
(2) after J by the local nerve impulse equations.

Phase (1) is given by the formula in Theorem 3, putting θ equal to the speed determined by the Corollary. Also put $y=0$, because the action potential is measured at a point, and starts at $t=0$.

The end of phase (1) triggers the beginning of phase (2). More precisely phase (2) is determined by the orbit on the slow manifold (see Figure 25) beginning at the *trigger point* J : $b=1 \cdot 2$ ($V=40$) ; $a=-0 \cdot 45$; $x=-1 \cdot 2$. The explanation of the trigger point is as follows. During phase (1) the propagation wave overrides the slow equation for b, and finishes at $V=40$ or $b=1 \cdot 2$. Meanwhile the slow equation for $\dot{a}$ has marginally reduced a from the equilibrium value $-0 \cdot 4$ to $-0 \cdot 45$. At the same time the fast equation has been acting to keep the state on the slow manifold, and so the value $x=-1 \cdot 2$ is obtained by solving $x^3 + ax + b = 0$.

Notice that during phase (1) four hidden events occur in rapid succession : first the pacemaker wave carries b past the threshold point $b=0 \cdot 1$; secondly the fast equation determines the action, which jumps the state from the upper surface to the lower surface of the slow manifold, rapidly reducing x ; thirdly as x crosses the value $x=-0 \cdot 5$ the membrane becomes permeable to sodium ; fourthly the rapid influx of sodium takes over as the main driving force in the slow equation for b during phase (2).

The calculated action potential is shown in Figure 26 and is drawn to the same scale as Figure 3 in section 1.1. Notice that it agrees remarkably well with the observed action potential in Figure 3b, and in fact has a slightly better shape than that given by the Hodgkin–Huxley theory in Figure 3a.

59

Heartbeat and nerve impulse

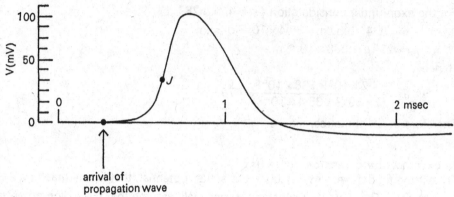

FIGURE 26
Calculated action potential

2.10. CONCLUSIONS AND EXPERIMENTS

We conclude by comparing our equations with the Hodgkin–Huxley equations [4, p. 518 and 5, p. 86]. Both are designed to fit the voltage clamp data, and both have similar time graphs for the potential and conductances, as shown by Figure 27. The graphs are drawn to the same scale, and it can be seen that in our theory the potential rises higher using smaller changes in conductance.

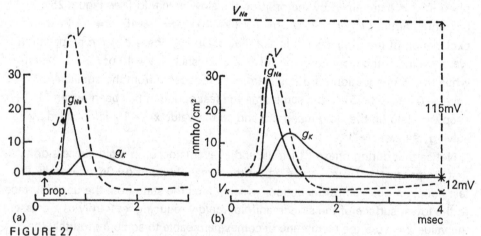

FIGURE 27
(a) Theoretical solution for propagated action potential and conductances.
(b) Analogous results for the Hodgkin–Huxley equations [5, p. 63]

The main points of difference between the two theories are as follows:

(1) Although we do not hold a very strong brief for our explicit equations, nevertheless they do belong to the general mathematical class of flows on the cusp catastrophe, which we deduced from the three dynamic qualities stated at

60

the beginning of the paper in section 1.1. Thom's deep uniqueness theorem for the cusp catastrophe shows that this class is not arbitrary but natural. We should expect this class to apply not only to the nerve impulse, but to any biological phenomena displaying the three dynamic qualities, such as spreading waves like epilepsy, migraine, or the Leão spreading depression, or discharge phenomena like reflexes, electric eels, or fireflies. Therefore our equations belong to a wider context.

(2) Since it is mathematically natural, and based on the simple dynamical qualities, the concept of representing the nerve impulse by a flow on the cusp catastrophe is well designed to withstand the inevitable buffeting by future discoveries about the underlying chemistry. By contrast the Hodgkin–Huxley equations are specifically based on the known chemistry, and may have to change fundamentally as further discoveries are made.

(3) Besides the electrical equation Hodgkin–Huxley use 9 *ad hoc* equations to fit the voltage clamp data, whereas we use only 1 *ad hoc* equation together with the 1 canonical fast equation.

(4) Hodgkin and Huxley [5, p. 86] assume that the change in potassium conductance is caused by a path for potassium being formed when 4 charged particles move to a certain region of the membrane under the influence of the electric field. As far as I know this 4-particle gate has not been confirmed by observation. If we trace the origin of the number 4 we find it came from data-fitting the delay at the beginning of the potassium conductance curves in Figure 24. Hodgkin and Huxley assume that the two permeabilities are in-dependent of one another, and are governed by linear equations. Their procedure is mathematically equivalent to the following : fit the end of the curve by exponential decay, and then cure the abrupt rise at the origin by squashing it down with a fourth power. Therefore, ignoring constants, they obtain $g_K = n^4$, and $n = 1 - e^{-t}$. The variable n is then given physical interpretation as the proba-bility that one of the gate particles is in the right place. Therefore n^4 is the probability that the 4-particle gate is in the right place.

By contrast we have a completely different explanation of the delay, because we assume the two permeabilities are interrelated. In essence we say that the potassium conductance cannot begin to rise until the sodium conductance has reached its peak. The situation is illustrated by Figure 23 : the delay occurs during the initial fast orbit FG, during most of which $\dot{a}$ is small negative, and $\dot{a}$ only becomes positive once the slow manifold is reached at G. Therefore we have no need to assert a 4-particle gate. Similar remarks apply to sodium.

61

(5) The discussion above prompts the question of what is the physical interpretation of our variables a, x. We have avoided the issue by calling them 'dynamical' variables, and relating them functionally to g_K, g_{Na}. Possibly they may also represent the concentration, or the production-rate, of some chemical that transports the ions. This interpretation would be consistent with the *cut-off* devices that we have built into the functional relations. For instance, the concentration of a has to reach the critical level, $a > 0.5$, before it can begin transporting potassium. Similarly $-x$ has to reach the critical level, $-x > 0.5$, before two particles of x can begin transporting a sodium ion. These are tentative remarks, but perhaps they may suffice to germinate more specific chemical hypotheses that can be tested experimentally. In fact perhaps the most useful purpose our equations can serve is to suggest alternative chemical hypotheses against which to test the Hodgkin–Huxley theory.

(6) An interesting feature of our equations is the very definite cut-off of sodium permeability, well away from the resting potential. For compare

Resting potential : $x = 1.07$

Sodium cut-off : $x = -0.5$.

Therefore in our model there is no inward sodium leak at, or near, the resting potential, nor is there experimental evidence for such. On the other hand there is a great deal of evidence for an outward sodium pump [5, chapter 6] caused by metabolism during the resting potential. It is certainly comforting in our model not to have to have an inward sodium leak and an outward sodium pump going on at the same time. More seriously, *the two surfaces of the cusp catastrophe may in fact reveal the fundamental difference between the two membrane states of sodium pump and sodium leak.* By contrast Hodgkin–Huxley [5, pp. 65, 73] assume a marginal inward sodium flow at the resting potential to explain the data in Figure 22. We explain this by the action of chlorine and other ions, rather than sodium, which mathematically gives rise to the saddle point shown in Figure 25. These features of sodium cut-off and saddle point are consequences of juggling with the position of the equilibrium point and functional relations, rather than being intrinsic to flows on the cusp catastrophe. However they might be a useful area for testing between the two models.

(7) The existence of the saddle point in our model has another interesting consequence, which might provide a fruitful source of experiments. If the constant 1.67 in the slow equation for $\dot{a}$ (*see* section 2.7) is increased to 1.7 then this has the effect of running the saddle and the sink together, and annihilating both. Therefore the stable equilibrium point representing the resting potential

would be replaced by a stable attracting closed orbit, containing the action potential. In other words, the neuron would start firing by itself periodically, without any trigger or propagation wave.

What would cause such a change in constant ? There might be a variety of causes because the same effect can be achieved by tinkering with other constants. Possibly use, or non-use, or external control of the neuron could change one or other of these constants. One can speculate on many uses for such a device, for instance :

(a) If the passage of one impulse increased the constant, then a volley of spikes would follow until fatigue reduced it again. In this way the neuron would act as amplifier for feeble messages.

(b) If the neuron had no dendritic input, then normal metabolism might increase the constant. Then the neuron would become a periodic self-firing neuron such as those found in the visual cortex, which possibly underlie the α-rhythm.

(c) The diffuse activating system of the reticular formation in the brain-stem might operate a control over the thalamus of this nature. If so this would help to explain the brain's facility in switching attention to, or from, a particular sense input.

(d) During sleep or dreaming, the brain-stem might operate a crude chemical control over the thalamus of this nature (such as the release of serotonin or noradrenaline by the Raphe system). This could explain the cut-off of sense inputs at thalamic level during sleep. The slowness of chemical control compared with the speed of neurological control in (c) above corresponds to the slowness of falling asleep and arousal compared with the speed of switching attention.

2.11. APPENDIX ON LAMARCKIAN EVOLUTION

We continue the digression of section 1.4. Recall the sequence of Examples 1–5 in sections 1.1 to 1.5 which we illustrate again in Figure 28.

Suppose that this sequence models the evolution of a linkage between two enzyme-systems of a cell. At the beginning (1) the two enzyme-systems are unlinked and are independently in homeostasis. At the end (5) the systems have become linked, and the state varies periodically round the van der Pol limit cycle. In other words the cell has developed a new biological clock.

Now biological clocks are of fundamental importance, not only for governing the timing of cell division, and so on, but also possibly for specifying position in the embryo (*see* recent experiments of Goodwin [13] and Wolpert [9]). In

Heartbeat and nerve impulse

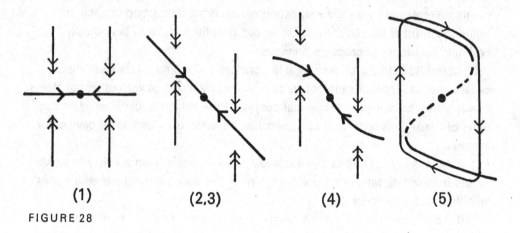

(1)　　　　　　**(2,3)**　　　　　　**(4)**　　　　　　**(5)**

FIGURE 28

fact the appearance of a new biological clock in the cell may lead to a new type of cell, and eventually to a new species. To emphasize the point we shall speak of the formation of the new biological clock as a *change in phenotype*. This is not such an exaggeration as it seems at first sight, for, as we shall see, the formation of the clock is governed by the cusp catastrophe, and generalizes at once to all higher dimensional catastrophes. Now any change in phenotype must depend upon some change in embryo, and Waddington [14] suggests that the mechanisms underlying embryonic developments are chreods, which Thom [1] describes mathematically in terms of catastrophes. Therefore any change in phenotype should have as underlying mathematics a strict generalization of the simple clock we are discussing here.

Summarizing, the passage from (1) to (5) represents both a change in genotype, because the enzyme-systems have to become linked, and a change in phenotype, because of the formation of a new clock. Therefore let us examine where and how the changes occur in the sequence. First we must recap the precise role that mathematics is playing in the discussion. Formally there are three steps involved. The first step from genotype to differential equations is *modelling*. The second step from the equations to the solution and to the attractors is *mathematics*. The third step from the attractors to the phenotype is *interpretation*. Behind the second step lie deep mathematical theorems such as the existence and uniqueness theorems of differential equations, theorems about the attractors of structurally stable systems, and Thom's classification theorem of elementary catastrophes. Therefore if the model is good, allowing identification of equations with genotype, and if the interpretation is good, allowing identification of solution with phenotype, then we should be able to call upon the mathematics to analyze the difference between changes in genotype and phenotype. The purpose of the appendix is to use the simple examples that we happen to have at hand to illustrate this point, as follows.

The identification of equations with genotype gives preferred coordinates, namely the coordinates x, b representing the two enzyme-systems. Therefore the step $(1) \to (2)$ represents a change in genotype because in (1) the slow manifold is a preferred coordinate axis, whereas in (2) it is not. As explained in section 1.4, the enzyme-systems in (2) are no longer independent. By (3) the enzyme-systems have become linked, and so the step $(2) \to (3)$ is another change in genotype. Both these steps may take a long time in evolutionary terms, and may only be the result of a large number of gene mutations and duplications. The step $(3) \to (4)$ is comparatively trivial because (3) is the linear approximation of (4), and therefore we assume that it represents no change in genotype. The step $(4) \to (5)$ is of a different nature, because, as pointed out in section 1.5, it can be achieved by a *continuous* change of parameter in the fast equation of the cusp catastrophe. Such a change is generic, and could be caused not by genetical changes but by a continuous change in environment, such as variation in temperature, climate, or food. Moreover such change could be very rapid in evolutionary terms.

Now look at the solutions : There is no change in phenotype from (1) to (4) because the only attractor is the stable equilibrium point. The only change in phenotype is the Hopf bifurcation $(4) \to (5)$, creating the clock.

Therefore we have a schema, which, to an observer, would appear remarkably Lamarckian, as shown in Figure 29.

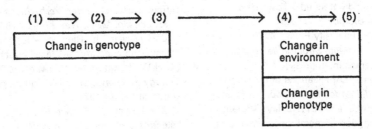

FIGURE 29

Of course it could happen only if the hidden random genetic mutations had stacked the cards ready beforehand, and afterwards the change would be reinforced by Darwinian natural selection on the phenotype. But it might help to explain why fossil records of phenotype can appear static over long periods, and then change relatively fast.

Similar arguments can apply to development as well as evolution. The continuous change $(4) \rightarrow (5)$ could be caused by diffusion between neighbouring cells coming into contact. For example, there may be a model of this nature associated with the development of mesoderm at the interface of contact between ectoderm and endoderm.

I am grateful to many people, particularly to René Thom and David Fowler for discussions about catastrophe theory, Boris Rybak and Ian Gray for discussions about the heart, Peter Buneman and Ted Woodcock for discussions about the nerve impulse, Francis Crick for suggesting at Serbelloni that catastrophe theory should be applied to the nerve impulse, and to Steve Smale, whose lectures [15] on electrical circuit theory at IMPA suggested how to put feedback on catastrophes into the classical framework of ordinary differential equations. I am indebted to Ted Woodcock [12] for checking the equations against the Hodgkin–Huxley data by computer. I should also like to acknowledge how many of the ideas sprang from A. L. Hodgkin's very stimulating book [5]. Finally may I apologize to my mathematical and biological readers, lest, by attempting to address the one, I inadvertently bore the other.

References

1. R. Thom, Topological models in biology, in (C. H. Waddington, ed.) *Towards a Theoretical Biology 3: Drafts* (Edinburgh University Press 1970).
2. R. Thom, *Stabilité Structurelle et Morphogénèse* (Benjamin, in press).
3. E. C. Zeeman, Applications of catastrophe theory, *Bull. London Math. Soc.* (In press.)
4. A. L. Hodgkin and A. F. Huxley, A quantitative description of membrane current and its application to conduction and excitation in nerve. *J. Physiol. 117* (1952) 500-44.
5. A. L. Hodgkin, *The Conduction of the Nervous Impulse* (Liverpool University Press 1964).
6. E. Hopf, Abzweigung einer periodischen Lösung von einer stationären Lösung eines Differentialsystems. *Ber. Verh. Sächs. Akad. Wiss. Leipzig. Math.-Nat. Kl. 95* (1943) 3-22.
7. D. H. Fowler, The Riemann–Hugoniot catastrophe and van der Waals' equation, in (C. H. Waddington, ed.) *Towards a Theoretical Biology 4: Essays*, pp. 1-7 (Edinburgh University Press 1972).
8. B. Rybak et J. J. Béchet, Recherches sur l'électromécanique cardiaque, *Pathologie-Biologie 9* (1961) 1861–71, 2035-54.
9. L. Wolpert, Positional information and the spatial patterns of cellular differentiation, *J. theoret. Biol. 25* (1969) 1-47.
10. Sampson Wright, *Applied Physiology* (Oxford Medical Publ. 1965, Eleventh edition).

E.C.Zeeman

11. A. L. Hodgkin, The ionic basis of electrical activity in nerve and muscle, *Biol. Rev. 26* (1951) 339-409
12. A. E. R. Woodcock, to appear.
13. B. Goodwin, *Temporal Organisations in Cells* (Academic Press : London 1963).
14. C. H. Waddington, *The Strategy of the Genes* (Allen and Unwin : London 1957).
15. S. Smale, On the mathematical foundation of electrical circuit theory (to appear).

Structuralism and biology

René Thom

Institut des Hautes Études, Bures-sur-Yvette

Can recent structuralist developments in anthropological sciences (such as
linguistics, ethnology, and so on) have a bearing on the methodology of biology ?
I believe this to be so ; after evaluating the proper task of structuralism, its aim,
its limits, we shall see how its prospective application to biology lies between
the old, Linnean type of biology, descriptive and taxonomic, and modern biology
(molecular biology, physiology, ecology) of a more reductionist, explanatory
nature.

STRUCTURALISM : ITS AIM AND METHODS

Morphological theory. We start from the following principle : any science is the
study of a morphology. By specifying certain sets of initial conditions (according
to a precise preparation procedure) the specialist of a given discipline may 'see'
some class of phenomena he particularly wants to study. This phenomenology is
always, in the final analysis, a spatio-temporal morphology. Of course, in many
sciences the substrate space of the morphology to be studied is no longer
physical space-time : it is a space of observables, of relevant parameters whose
precise nature may be difficult to specify (as for social sciences for instance).
Nevertheless, in all cases, the morphology is deduced from observation in
physical space-time, and it inherits out of it a topology which always plays some
role in the definition of the substrate space which supports the morphology to be
studied. As already defined in a previous article, we shall describe a morphology
by a set of qualitative discontinuities in the substrate space (the so-called
catastrophe set) [1]. This set is the complement of the set of 'regular points',
where the qualitative appearance of the process varies continuously.

In most cases, the given morphology exhibits properties of structural stability.
Following experimental perturbation of the local initial data (or when this is not
feasible, as for stellar objects, or sociological phenomena, by repeated observations), some local morphological accidents may remain topologically the same
under sufficiently small perturbation ; we say, in that case, that they are described
by a 'chreod', using Waddington's terminology. Each chreod is described by a
local model, giving the topological type of the corresponding catastrophe set. In
many cases, it is possible to construct a finite dictionary of all chreods present in

68

our given morphology : the catastrophe set of any experience is then covered by a finite number of such elementary chreods. The macroscopic morphology of everyday life satisfies this requirement, as we can in fact describe all the usual objects we meet by the words of a finite dictionary.

In general, the morphology of a given experimental situation varies with a variation of the initial data ; if such a morphology is unique up to isomorphism, then it can be described by a unique maximal chreod. Such (or almost such) is the situation of a discipline like Human Anatomy, which describes the internal structure of the human body, any accidental variations of its morphology being rejected as 'pathological'. But, in general, this is not the case, and we are given an 'ensemble' of morphologies, obtained by properly varying the initial data. The initial task of the experimentalist is to register the totality of this ensemble— what in linguistics is called the 'corpus' of a given language.

In such a 'corpus', some aggregations of chreods appear more frequently than others. It is sometimes possible to prescribe a subclass of initial data which makes such an aggregation stable. If this is so, we say that such an aggregation of chreods is a *conditional chreod*. For instance, the presence of a living being in a domain of space requires the presence of another living being, its parent, in the initial data (*Omne vivum ex ovo*). Almost any morphological discipline has as primary task : the description (and the preparation) of all present conditional chreods. In linguistics, for instance, phonetic study of the spoken language shows that this sound morphology can be decomposed into a finite set of irreducible elements, the *phonemes*. But the true object of linguistics is the study of such conditional chreods as words or sentences, aggregation of phonemes whose emission can be made stable in proper environmental situations. In this way, the notion of conditional chreod allows us to take care of the notion of *order* and *hierarchy*.

▶ *The notion of a system.* Here we define a real system (in opposition to the formal systems of mathematicians and logicians) as an open connected set of regular points in substrate space whose closure is a conditional chreod. The decomposition of such a domain into elementary chreods is called the *structure* of the system. In most cases, the structure of a system is time-invariant, but in some cases it may allow some variation for occasional regulation—or at most a slow change within the spatio-temporal continuity in substrate space which does not essentially affect the stability. 'A man who loses his hair by age remains nevertheless a man.' Qualitative continuity in substrate space is a sufficient condition for permanence of the system.

With such a definition, a 'word' is a real system of linguistic morphology. Both the words in the language and the object they describe for the most part satisfy our connexity requirement. In such a case, most objects are topologically balls. An electric circuit is topologically a circle; two electric circuits are disjoint if their interactions by electromagnetic induction is negligible; if, on the contrary, they are inductively coupled by two coils, they form a single system.

This definition of a system may seem surprising; generally a system is defined as a set of interacting elements. The morphological definition given here has two advantages:

(1) it does not impose the choice of elements, which may be arbitrary;

(2) conversely, if elements are given in a morphology, we escape answering the tricky question of which sets of elements form a system, and which not.

▶ *Real system and mathematical objects.* We have seen that the morphological repetition of locally structurally stable accidents is the basis of any morphological, structural analysis. But mathematical objects often have the same property: for instance, in a polyhedron P with a symmetry group G, if x' is a vertex of the orbit $G \cdot x$ of x in G, then the local structure of P at x' is isomorphic to the local structure at x. The same holds for the smooth action of a compact Lie group G on a compact manifold M, and then the possible local types of actions are finite in number. Another, less well-known source of morphological repetition in mathematical objects is the 'genericity' assumption. For instance, all smooth real valued functions on $\mathscr{R}^n$ belonging to a suitable open dense subset in the C^2 topology

$$\left(\text{with distance } \left[|f| + \left| \frac{\partial f}{\partial x} \right| + \left| \frac{\partial^2 f}{\partial x \, \partial x} \right| \right] \right)$$

have, as singularities, only non-degenerate critical quadratic points, that is, points x with $df(x) = 0$ and

$$\left| \frac{\partial^2 f}{\partial x_i \, \partial x_j} \right| \neq 0 \quad .$$

One can say that the long-range aim of any structuralist theory is to explain the observed empirical morphological repetition by associating to it, as model, a mathematical object of isomorphic structure. So the task of any structuralist theory is:

(1) to form a finite lexicon of elementary chreods g_i;

(2) to build experimentally the 'corpus' of the empirical morphology;

(3) to define the 'conditional chreods', objects of the theory;

(4) to describe the internal structure of a conditional (or elementary) chreod by associating a mathematical object to it, whose internal structure is isomorphic to the structure of the chreod.

70

More precisely, if a conditional chreod c has for support some domain V in the substrate space, and if the mathematical object we wish to associate to c is a geometric object lying in a space U, then c is induced by a mapping $\varphi : V \rightarrow U$. The recent use of structuralist methods have led to a relative success in only a few cases (phonetics, structural linguistics, ethnology). Nevertheless, the epistemological importance of such methods should not be underestimated. From a qualitative point of view, they represent an analogous position to the famous *Hypotheses non fingo* of Newton. In the structuralist viewpoint, one does not try to explain a morphology by reduction to elements borrowed from another theory—supposedly more elementary or fundamental—as one might try to explain biology by physics and/or chemistry, or sociology by psychology or biology ; one only tries to improve the description of the empirical morphology by exhibiting its regularities, its hidden symmetries, by showing its internal unity through a formal mathematical model which can be generated axiomatically. In that respect 'structuralism' is a modest theory, as its only purpose is to improve description. Nevertheless, because of its extreme generality, this attempt raises three questions.

(1) The first deals with the uniqueness of the proposed mathematical model. As is well known through examples, the decomposition of a morphology in elementary chreods is to some extent arbitrary (not uniquely defined). May we hope that if an empirical morphology M admits two different mathematical models F, F', there exists a finer one F'' which generates both of them ?

(2) What is the mathematical nature of the objects F to be identified with the empirical morphology ? A polyhedron with a symmetry group ? A pure algebraic structure (group, ring, field) provided with a convenient topology ? Or a graph given without any motivation ? We have so far no idea how to answer this question. In anthropological sciences, the mathematical structures used up to now are fairly rudimentary : practically the dichotomic opposition, expressed by the Z_2 group acting on the real axis.

(3) To be complete, we have to quote the Baconian objection to those scientists who are fascinated by the physical model : while a physical law, like the gravitation law, permits an infinity of experimental verifications, a purely qualitative model, like the structuralist construction, is not susceptible to experimental control : there is nothing to do but to check, once and for all, the agreement of the model with experiment. Some conclude from this that structuralism is nothing but an academic entertainment, not worthy of a true scientist's

attention. Although I do not share this obscurantist opinion, I nevertheless believe that 'structuralism' should motivate its models, by reintroducing dynamics, time and evolution of structures. By doing so, it would be theoretically possible to perform qualitative experiments on an empirical morphology, by properly choosing perturbations of the initial data (at least, of course, for disciplines for which experiment is feasible ; the status of the other disciplines, like paleontology, geology, sociology, and so on, or morphology of stellar objects, with no possibility of experiment, will always remain somewhat dubious).

▶ *The epistemological problem*. The agreement, often observed in numerous disciplines of the animate and the inanimate world, between an empirical morphology and a mathematical structure raises a classical problem of epistemology. Three types of answer may be given.

(1) The first attributes the agreement to a pre-established harmony between mathematics and reality. This is the Platonic (more precisely Pythagorean) viewpoint : 'God ever geometrizes'.

(2) The second appeals to the principle that the phenomena are governed by a condition of local equilibrium, or more precisely are the solution of an extremality problem.

(3) The third—the one I am advocating—explains the origin of mathematical structure (and its morphological repetitions) by a genericity assumption. In every circumstance, *nature realizes the local morphology which is the least complex possible with respect to the given local initial data.*

The first answer is pure metaphysics. The second is the only one which may be considered as strictly scientific, as it can sometimes be verified on quantitative models. For instance, according to Prigogine [2], Bénard's phenomenon (that is, the formation of a hexagonal pattern of convection cells in a heated fluid), can be explained by a local minimum of entropy production. The third answer is intermediate between science and metaphysics ; it has the advantage of taking a much more flexible viewpoint than (2), which is more global. The theory of catastrophes explains the repetition of local morphological accidents by the isomorphism of local conflicting dynamical situations which generate them. Answers (2) and (3) are in fact not incompatible. For instance, the solution of an extremal problem (like Plateau's problem : to find a minimal surface with given boundary) will present 'generic' singularities for almost any given boundary data. Answer (3) does not take an *a priori* position on the problem of determinism. In the third viewpoint, determinism is not an *a priori* postulate, but rather the ultimate aim of the theory.

72

R. Thom

Two basic examples. In our morphological approach, a system is in the first place a connected region of the substrate space. Hence one of the first properties to consider when dealing with a system is its spatial regulation, that is, the behaviour of its boundary when subjected to external perturbations. We want to consider two systems, which are basic (and fundamentally opposed) examples of what can happen in this direction.

The first is the physical system : perfect gas. Morphologically it has the following properties :

(1) the system does not regulate its boundary : a gas takes the shape of its container ;

(2) there are no intermediate morphologies (chreods) between the total system and the individual molecules. Moreover, for a chemically pure gas, all molecules are alike.

For such a system, its macroscopic behaviour may be deduced from postulated properties of the interaction between nearby molecules. Knowledge of the interaction potential, under very weak restrictions—like dominance of the repulsive part on the attractive, and sufficiently rapid vanishing at infinity—is enough to obtain the state equation law of the gas. Removing a molecule from such a system has no effect on the global behaviour. There is an enormous *morphological redundancy.*

Let us consider, at the opposite extreme, a man-made machine, a clock for instance. It has the following properties.

(1) The system regulates its own boundary. In fact, uniform rotation of the hands is the machine's function.

(2) Between the global system and the molecular level, there is an intermediate level defined by the parts of the mechanism (springs, cogwheels, and so on).

(3) These parts present some morphological repetition : most of the cogwheels are isomorphic.

(4) On the contrary, removing only one part suffices to stop the mechanism, there is *morphological repetition,* but not *morphological redundancy.*

These two systems are the two typical examples between which theoretical biology has never ceased to oscillate. The first model, the model of statistical mechanics, has the advantage of being completely quantitative and computable. But it is an ideal situation, valid only asymptotically for real gases. (Recall that, up to now, there is no molecular theory of phase transitions.) The absence of regulation of spatial form, the non-existence of an intermediate morphological

73

level, of a structure, suffices to show the huge gap which separates this system from living beings. The second model is that of Descartes' 'animaux machines', which has been modernized under the name of cybernetics. The analogy is relatively valid for animals, where, at the organ level, there is seldom morphological redundancy. For plants, however, morphological redundancy is the rule for all organs (leaves, flowers, roots) except the central stem. Hence the idea of 'vegetable machines' is somewhat strange.

Caught between these two extremes, the atomic or reductionist model on one side, the cybernetic one on the other, both obviously insufficient, how can theoretical biology get out of this 'stalemate'?

The only hope of getting out of the difficulty is to recognize that there is no gap between the two kinds of systems, and that both may be embedded in a continuous family which contains both of them. This compels us to abandon, at least provisionally, the main merits of the two models : the quantitative, computable aspect of the first, the diagrammatic cybernetic of the second. We have to appeal to the only common factor of the two systems, that is, their spatial extension and structure, their morphology. Moreover for a man-made system like a machine, it is compulsory to take its genesis into account, that is, the psychological processes which take place in the mind of its designer, when he starts building it. It may seem strange, or far-fetched, to assimilate a psychological process to a physical one. To justify this approach, we shall digress somewhat on the technological problem in general.

▶ *The technological problem.* The purpose of any tool, of any machine, is to realize a task that man could do, but which would be difficult, dangerous, or tiring. As has often been repeated—from Bergson to MacLuhan—the main function of a machine is to complement and amplify our means of biological self-regulation. In particular, regulatory catastrophes involved in feeding, in moving our own body or our belongings are the main purposes of our machines. Hence, any machine has an 'active site', a part which is in contact with the object to be moved, and which has to describe a specific trajectory (or set of trajectories) X ; these trajectories are described either in ordinary space, or in the product of ordinary space by a space of physical or physiological parameters (for instance : temperature for a heating device).

In general, the designer already has at his disposal 'elementary parts' P_i, each of which has a configuration space X_i of the same qualitative nature as the space X to be realized. For instance, for a mechanical machine like a clock, the space X is the circle described by the extremity of the hands, an elementary part is a cogwheel P, whose configuration space is also the circle S_1. Moreover, between these elementary parts, there are standard constraints which connect a given space X_i of P_i to the space X_j of P_j. For instance, if the cogwheel P_1 has N_1 teeth geared to P_2 (with N_2 teeth), then between the angular coordinates θ_1, θ_2 of P_1, P_2 respectively, there is the relation $\theta_1/N_1 = \theta_2/N_2$, through which both configuration spaces X_1 and X_2 are identified. Let us neglect, for the time being, the obviously fundamental problem of the energy supply : then the problem to be solved is to get the space X as the quotient of spaces $\Pi(X_i)$ of the parts P_i by the relations r_{ij} defined by standard constraints between X_i and X_j.

This problem is in general largely indeterminate : hence the machine can be constructed in many non-isomorphic ways. Only some spatial constraints have to be respected : the ones concerning the active site of the machine, and the energy source. But the spatial disposition of intermediate parts is largely arbitrary. The nature of elementary parts P_i and of the constraints r_{ij} obviously depend on the qualitative nature of the space X to be realized. For instance, in the construction of an electronic circuit, the elementary parts are conductors, resistances, coils, transistors, batteries, and so on, and the constraints r_{ij} are the contacts between these parts which identify the electric potential at these points. In such a case, the spatial ordering of the parts on the circuit board is irrelevant : only the topology of the connections to be realized in the diagram matters.

This fact is at the origin of a widespread illusion : namely the idea that any regulatory mechanism can be described by a diagram of feed-back type (graph with arrows), possibly generalized. Applying this analogy to biological systems involves two misconceptions.

The first is that it overlooks the fact that in living organisms almost all regulatory fields or activities (if not all of them) *depend on continuous parameters*. For instance the 'walking "field" in man can be adapted to sandy or slippery ground, to a slope', moreover, the man may decide to run, instead of walking. How could such a continuous (or even discontinuous) variation of the motor fields be described by a discrete graph ? The only possible description involves necessarily a space of relevant parameters to describe the external stimuli, with continuous catastrophe hypersurfaces in it. So, if only to describe the complete correspondence between stimuli and responses, a continuous topological model is necessary, and the cybernetic diagrammatic description totally insufficient.

Secondly, a technical diagram for building a machine makes sense only in relation to a given technical situation, a 'presupposed' technology, which associates to any symbol of the diagram an elementary organ P_i. The designer may take such a situation for granted : he does not have to worry about energy supply, about the obedience or capability of his workers, nor about providing raw materials for the parts. This may not be true for the biological machinery. One might ask, in particular, if it is really necessary for a living organism to have a programming organ, a role classically attributed to the genome by biologists.

▶ *The myth of the programming organ.* Let us consider a machine like a washing-machine running on electricity. Here, the space X described by the active site of the machine is a qualitative one (soaking, washing, rinsing, and so on). These changes are controlled to take place at the right time according to a program tape, which the machine itself unrolls. This tape supports an internal spatial morphology isomorphic to the space X itself, described by the active site of the machine. Generalizing this analogy, we shall say that if in a system S whose global temporal evolution admits a spatio-temporal morphology M, some sub-system S' of S is a program organ if (1) its internal morphology is isomorphic to M, (2) removing S' implies the complete stoppage of the system.

How do such criteria apply to the genome ? There is no obvious relation between the internal morphology of the chromosome (as either a DNA molecule or a more complex protein–DNA complex) and the global morphology of the living organism. According to B. Goodwin [3], an activity wave runs through the circular chromosome of *E. coli* during the mitotic cycle : it would be very interesting to know if the same is true for the phage's chromosome in vegetative reproduction or to the Metazoa chromosome in gametogenesis. However the second criterion of the last paragraph seems to apply to the DNA, at least in as much as we are interested in reproductive activities (some cells may live a very long time without DNA on a purely vegetative basis).

Generally, we shall entitle *programming organ* any part P_i such that in the canonical map $\Pi(X_i) \rightarrow X$, we have a lifting $X \rightarrow X_i$ of q. With this definition, there are a lot of programming organs (for instance, any cogwheel in a clock) : *the non-uniqueness of the programming organ is the general rule in machines and biological systems*. We should first observe that in animals only some regulating functional activities are highly canalized ; it would be preposterous to speak of a global program : the variability of responses to stimuli, which are themselves unforeseeable in their succession, and even in their nature, offers the organism its only chance of adapting itself to an ever-changing environment. We can speak of programs only for relatively stereotyped regulatory activities, those which are usually called *functions*. In that case functions have not only one but several programming organs, with a very great morphological repetition (and even redundancy).

▶ *The constitution of phases and the field hypothesis.* Let us go back to the real gas system ; its morphological analysis showed that there is only one elementary chreod, the 'molecule'. But these elements may group themselves in space according to specific stable patterns, the 'phases'. Under very general assumptions (rapid decrease of the interaction potential with the distance), all possible local organization of molecules may be parameterized by points of an auxiliary space F, taken as the fibre space of 'internal states' of a fibre space over the substrate space U. The computation of the potential energy around a point x of U gives rise to a potential function $V : F \Rightarrow R$. The phases of the stable organizations of molecules correspond to minima of V. The partition of substrate space into phases, and the limit surfaces of phase transitions, are defined by the way V_x varies as a function of x. This is usually called a 'surface' effect. In classical statistical mechanics, the aim of the theory is to show that surface effects can be neglected when the walls of the container go to infinity : in biology, on the contrary, these surface effects are essential, as they depend on the global configuration of the system, the regulative properties of which they express.

From this point of view, we have a relatively strict isomorphism between the major biological phases (that is, the main cellular differentiations) and the major regulatory activities of living organisms ; in particular they correspond to the qualitative types of X spaces which intervene in biological regulation. For instance spatial regulation involves : (1) static regulation, for example, an extremely resilient material provided by bone tissue ; (2) cinematic spatial regulation, for example, motion, provided by muscle tissue. Generally the nuclei of a new phase have a tendency to get organized according to a specific internal determinism, which can be affected by local factors (surface effects !). Hence we get organs P_i, with configuration space X_i, and standard constraints r_{ij} between them (for instance, attaching a muscle to a bone by a tendon). Thus the technological problem and the problem of organogenesis are exactly alike : in both cases the problem is to form some global regulatory space X as a quotient of a product of organ spaces X_i, through standard constraints r_{ij}. One could think of their solutions as being of the same nature : the technological solution appears in the designer's mind as a result of a competition process between several conceivable solutions ; in the same way, the solutions arrived at in life by organogenesis are the result of competition processes, both internal and external : the external one is natural selection ; the internal one is realized by a metabolic simulation of various solutions, most of which fail to create the corresponding morphological structures.

This general scheme shows that a knowledge of the fine structure, molecules for a fluid, cells for an animal, is practically irrelevant for understanding the global regulatory figure (the 'logos' as I propose to call it) of the total system. For instance, the final structure of a theory like Fluid Mechanics does not depend on whether one takes as the basic concept molecules or a continuous fluid ; in the same way, our model leads us in all cases to consider the local state of the system as defined by a 'field' ; a local potential on a fibre space F. This theoretical situation is similar to what occurs in elementary particle theory, where one also forgets the particle to consider a local field only (of a different mathematical nature, of course). Is there any possibility that one might have to 'cross levels' and explain a gross morphology by details of the fine structure ?

▶ *Old and new biology*. The branches of traditional biology each dealt with a specific hierarchical level : anatomy, physiology, for the global organism ; cytology at the cell level ; molecular biochemistry at the molecular level. These disciplines are fundamentally descriptive and taxonomic. With genetics the situation is different ; genetics is an example of a 'cross level' discipline, as it hopes to describe correlations between some macroscopic indeterminacies at the gross level (organism) and fine structure details of the genetic material at the molecular level. These are still very mysterious. The 'genetic code' may explain how DNA determines the primary structure of a protein, but not how to relate the primary structure of the protein to its tertiary (or quaternary) structure, hence to understand its regulatory activities. This is a problem which has still evaded all attempts at a proper formalization and which seems very difficult to solve. It is not impossible that the genetic code may in fact be a 'ritualization', which occurred relatively late in evolution, of molecular mechanisms which were at the beginning much more flexible and variable. This is why the conceptual problem of 'cross level' regulations is still practically open.

▶ *Cross level regulations*. In the catastrophe model, the totality of regulatory mechanisms of a system (animate or inanimate) is described by a regulatory figure (or 'logos') which can be thought of as a volcano crater (the Gamow crater of a nucleus in atomic physics) : here sometimes the walls of the crater form overhang, and this gives rise to discontinuities in the homeostatic behaviour of the system. All this figure may be supposed embedded in the universal unfolding of a unique singularity, the 'germinal point', which is the organizing centre of the whole structure. In Metazoa, the system is sometimes reduced to a single cell (spore or egg). Hence we have to admit that there exists an oscillation between the two levels : cellular and multicellular ; this fact implies that the regulation figures of the organism and of the cell are vaguely isomorphic, at least, that each one may be continuously deformed into the other.

Such a hypothesis is relatively easy to satisfy in a geometric model with local fields of potential function type. If the regulation of a cell is described by a potential function $V = R^n \to R$, the space of potentials $\mathscr{L}(R^n, R)$ is the space F of all possible metabolic states of the cell. In an organism which is a three-dimensional ball B^3, the field is described by a continuous (in principle) mapping $g : B^3 \to \mathscr{L}(R^n R)$, hence the space of the fields is the function space $\mathscr{L}(B^3 \times R^n . R)$. But this space is isomorphic to the preceding one for large n.

While gametogenesis has to be interpreted, at organism level, as a local reverting to the organizing centre of the whole structure, the same process, taken at cell level, has to be considered as an unfolding. To explain that, we have to discuss the linguistics–genetics analogy.

▶ *Linguistics–genetics analogy.* The linear structure of the genomic material has suggested naturally the analogy with a written text : but there is another analogy between embryology and linguistics which is less well known, and worth discussing.

Let us recall that embryological development (in Vertebrates, for instance) can be described by a graph, a tree (the epigenetic landscape). From the initial vertex, the egg, three branches set out describing the main embryological sheets : ectoderm, mesoderm, endoderm. Development proceeds by further ramification of these branches

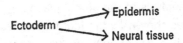

In the 'hydraulic' model of the epigenetic landscape, these branches are represented by rivers descending to the germinal point, which is the absolute sink.

In the linguistic theory known as *generative grammar* we meet exactly with the same structure. A sentence like

the cat ate the mouse

is described syntactically by a graph, a tree, the initial vertex is the P vertex (meaning sentence) : it splits into three branches, which end into terminal symbols (that is) words, as shown in Figure 1.

There is a certain isomorphism between these two graphs, even on the functional meaning level : ectoderm corresponds to the subject, which it limits

79

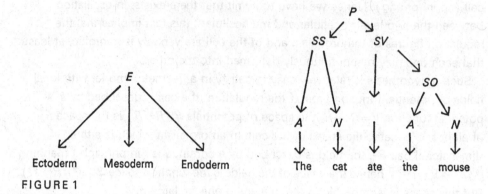

FIGURE 1

spatially (and in the stimuli space, as neural tissue) ; the mesoderm corresponds to the verb, as it is specialized in movement, and action ; endoderm corresponds to the prey, the grammatical object.

This analogy, however striking it may be, is nevertheless in contradiction with the general scheme of the Egg—Message analogy, shown in Table 1.

TABLE 1

Biology	Adult	Gametogenesis	Egg	Development	Adult
Linguistic	Meaning	Emission	Speech	Reception	Meaning

To reconstruct the analogy, we have to take into account the fact that *emission* of speech starts with an analysis of the global meaning : this operation requires a lot of external energy (as shown by the fact that aphasic troubles are more likely to impair emission than reception) : in this process elements are brutally separated, exactly as chemical analysis of a compound substance requires external heating or energy. It is not unlikely that gametogenesis uses the same kind of analysis : the coupled elements of the regulatory structure (that is, the controlled genes of the differentiating gametocytes) are isolated by 'heating', and then solidified as 'prepared' genes in the gametic genome. In that respect, gametogenesis is an unfolding at the cellular level : the egg is probably the most complicated of all cells in the organisms of a living species. The polarization of the egg along the animal-vegetative gradient is analogous to the order : subject-verb-object in the spoken transitive sentence. Quite probably, the regulatory figure of a species as a whole arises from a coupling, a conflict, between several

80

antagonistic regulatory systems. After the analysis realized in gametogenesis, these subsystems are separated and isolated as molecular structures of the genome. In the metazoa these structures may exhibit repetition or redundancy, contrary to what occurs in bacteria, like *E. coli*, where it is admitted that only one copy of any gene exists on the chromosome, the number of chromosomes in a cell being in general two or three.

In that respect, if embryological development starts as an unfolding, this unfolding continues the molecular unfolding of the egg along the animal-vegetative gradient. The analogy cannot be pushed further, even at the neurulation stage. At the end of development, despite the morphological diversity of cellular differentiation, we get the fundamental functional, self-regulatory unity of the organism, described by a connected regulatory figure, homologous to the global meaning of the sentence understood by the auditor.

The genetics–linguistics analogy, however seducing it may be, is not to be taken too literally. Biological structures are three-dimensional and not uni-dimensional, like speech or writing. The genome's uni-dimensionality is probably more due to a self-replication constraint than to a communication constraint. Personally I believe that the analogy has more validity in the direction opposite to the usual one. Language was created in man, when the 'genetic field' invaded the 'conceptual field' : concepts in the mind became capable of reproducing themselves, and of constructing gametes of their own : the words. Summarizing : linguistics may be explained by an extension of genetic mechanisms, rather than the converse [cf. 3].

▶ *Effect of the gross structure on the finer level.* If through genetics we understand how the fine structure may affect gross behaviour (and this, probably, only at the most indeterminate phases of the gross evolution, in the 'generalized catastrophes' occurring in development), we have also a reverse effect : the gross structure may act at the local level. As already explained, the 'stable biological phases', stable cellular differentiations, correspond to thresholds on the Gamow rim of the regulation figure : when two nearby cells have to meet in a common tissue, they enter into strong resonance. Their representation points have to come very close, and this requires some lowering of the limiting threshold in at least one specific regulatory direction. 'Phases' correspond to a 'threshold stabilization' on the Gamow wall of the regulatory figure. If the gross organism has itself a regulatory figure relatively isomorphic to the cell, starting a main regulatory reflex will necessitate triggering the activity of the corresponding 'phase', whose threshold is at the source of the reflex on the Gamow rim. For this reason,

Structuralism and biology

collective life often imposes (only individuals of the germinal line may be excluded) a deformation of the regulatory figure of the individual. This deformation, by lowering a threshold of the wall, diminishes the stability of the individual element. In fact, in a living organism, some cells may eventually receive a 'suicide order' to insure spatial or physiological regulation of the organism (cf. epidermal cells, hematocytes, and so on). In human societies, this deformation is called 'ethics'; as is well known, the egoism of a society is based on the altruism of its low-ranking members.

The author wishes to thank Mr M. Reid, who helped him to translate this article.

Notes and References

1. R. Thom, Topological models in biology, in (C. H. Waddington, ed.) *Towards a Theoretical Biology 3: Drafts* pp. 89-116 (Edinburgh University Press 1970).
2. K. Prigogine, *Introduction to the Thermodynamics of Irreversible Processes* (Thomas Springfield 1955).
3. B. Goodwin, *The Temporal Organization of Cells* (Academic Press : London 1963).
4. In the 'genetics-linguistics' analogy, the homologue of the nucleotide sequence of DNA is the phonemic structure of speech (or the alphabetic structure of writing). It is here, at this lowest level, that the Saussurian 'arbitrariness of sign' is most conspicuous, here the effect of random fluctuations in the past history is most important. At the higher level of syntax, however, the word order is much less arbitrary, as it manifests a relatively tight connection with meaning [*see* R. Thom, *Topologie et Linguistique: Essays on Topology*, pp. 226-48. (Springer Ed. 1970)]. This is why the present contention of molecular biology, that the genetic code explains morphogenesis, is so ill founded ; it amounts to saying that deciphering the alphabet of an unknown language suffices to understand it.

The concept of positional information and pattern formation

L. Wolpert
The Middlesex Hospital Medical School

The expression of genetic information in terms of pattern and form is a central problem, not only for developmental biologists, but for biology as a whole. The translation of genetic information into shapes and patterns is what links genetics to morphology—genotype to phenotype—and must have crucial consequences for a variety of central biological issues from evolution to learning. It is, for example, becoming increasingly clear that if we understood the mechanism whereby nerve connections were formed we might have a much better chance of understanding the neural mechanisms involved in learning [1].

Does the genetic material provide a description of the adult ? What, one may ask, are the genes for leg formation in tetrapods, and how do they make a leg ? Or, what are the genes for gastrulation ? A current fashion in molecular biology is to suggest, either explicitly or implicitly, that the answers to such problems will come from deeper and deeper molecular probings. Characterize the RNA and proteins and the form will look after itself, or at least be immediately explicable [2]. Such a view suggests that if we understood cytodifferentiation or molecular differentiation then pattern would be explicable. I wish to take a rather different view and would suggest that the development of form and pattern, while related to molecular differentiation, can be viewed in their own right. Moreover, the rules, laws, or principles for the expression of genetic information in terms of pattern and form will be as general, universal, elegant, and simple as those which now apply to molecular genetics [3, 4]. Again in contrast to the prevailing ethos, I would suggest that the solution to the problems must initially be sought not at the molecular level but at the cellular level, since until the cellular basis is understood, the correct molecular questions cannot be posed. Here I will briefly consider some general aspects of pattern formation in terms of positional information. (For a more extended discussion *see* Wolpert [5].)

▶ *Pattern and form.* A useful but by no means absolute distinction between form and pattern [cf. 6] is that form involves cell movement and changes in shape, its genesis requiring an understanding of the forces involved, whereas pattern does not involve changes in shape or cell movement, but rather the specification of spatial differences. Thus, the distinction rests in the nature of the cellular

83

processes involved. Pattern formation will usually precede cell movement since it is necessary to specify which cells will move, and where. In very general terms, the pattern problem may be stated as follows : given an ensemble of more or less identical cells, how can states be assigned to these cells such that when they undergo molecular differentiation the cells will form a well-defined spatial pattern. The difference between molecular or cytodifferentiation and spatial specification is a crucial one. Molecular differentiation can be viewed in terms of the control of synthesis of specific macromolecules, particularly luxury molecules, to use Holtzer's evocative terminology [7]. The process of pattern formation on the other hand must be viewed in terms of the spatial organization of such activities. For example, cytodifferentiation in a tetrapod arm and leg is probably very similar with respect to early muscle and cartilage development: the difference between an arm and leg lies in the spatial arrangement of these processes. A suggested solution to the pattern problem is that the cells are assigned positional information which effectively gives them their position in a coordinate system and this positional information is used to determine the cells' molecular or cytodifferentiation [3].

An example of the relationship between pattern and form comes from sea urchin gastrulation which involves pseudopodal activity by mesenchymal cells providing the cellular forces for invagination [8, 9]. The form resulting is due to the deformation of the cell sheet. The pattern aspect of this problem is the assignation to a particular group of cells of the cellular property of pseudopodal activity.

Our studies on the early development of form in the sea urchin embryo [8, 9] and other studies [see 10] suggest that the cellular forces responsible for cell movement and the generation of form are relatively few in number and are repeatedly used in a variety of animals and processes. In particular, localized cellular contractions whether in pseudopods or near the cell surface play a central role. This type of cellular force has been invoked for a variety of changes in form in the sea urchin, but in other systems such as neurulation [11] recent studies by Wessells *et al.* [12] have emphasized the common role of contractile filaments in a wide variety of such processes. It presents no conceptual problem to discuss contractile processes near the cell surface in terms of gene action, however ignorant we remain of the mechanisms [13]. It is worth recalling that virally transformed cells often involve changes in cell surface properties and cell motility and it is possible that from such studies one might obtain the best clues as to how genes affect surfaces and motile processes [for example, 14].

84

L. Wolpert

One important aspect of such mechanisms for generating form by cellular
forces is that the specification of how to bring about the change in form is almost
always very much easier than describing the changes that result. It is, for example,
easy to specify a mechanism for gastrulation in sea urchin embryo, but rather
difficult to describe the final result. We have used this point to emphasize that
the genetic information in the egg does not contain a description of the adult
but a set of instructions on how to make it, and that this set of instructions is
much simpler than any description [15].

▶ *Pattern regulation*. The formulation of the pattern problem given above is somewhat misleading, starting as it does with the cell ensemble, since most real
biological problems deal with situations in which the ensemble is derived by cell
division from a single cell, the egg. In many situations pattern is specified during
or following subdivision of the egg. Viewed in these terms one can distinguish
between two extreme situations. In the one there is very little interaction between
the cells and the differences that arise between the cells reflects differences
already existing within the egg. In the other, the differences arise due to intercellular signalling. In the so-called mosaic case, removal of a small piece of the
system results in a localized defect. By contrast in regulative systems there is
usually no defect or, as Sander has pointed out [16], a much greater and widespread disturbance occurs. In general, most systems show both aspects but at
different times [17, 18]. Another way of looking at the problem is to consider
it in terms of presumptive regions and potentialities. Presumptive regions are
classically defined as indicating which cells will *normally* develop into a particular
structure such as eye or limb. Following Driesch [19] and Spiegelman [20] one
might formulate a Driesch–Spiegelman law along the following lines : the amount
of material capable of developing into a particular region is always greater than
the presumptive region. Since the potentiality is always greater than the expression, some mechanism for restriction of the expression is required. It will be important to know if the law is universally true, particularly in mosaic eggs.

The situations in which cell-to-cell communication is clearly dominant are
those situations in which regulation occurs, either during development as in sea
urchin or amphibian development, or in regeneration as in hydra or amphibian
limbs. These can be illustrated by the French Flag problem [21]. This problem is
concerned with the necessary properties of cells in a line, and the intercellular
communication between them, such that if each cell has three possibilities for
molecular differentiation—blue, white, and red—the system always forms the
French Flag irrespective of the number of units or which parts are removed. In

terms of positional information, the solution is that each cell is assigned, by appropriate signals, a positional value which specifies its position with respect to the two ends. The cells interpret this positional value by turning on the genes for blue, white, or red, and so form the French Flag. The positional signal may involve, for example, a diffusing substance, or might be the time difference between two periodic signals. In both cases they would specify a positional value which may be looked upon as a cellular property which was graded from one end of the line to the other. It could be a gradient in one, or two, substances or some more complex cellular property [see 5]. The crucial feature is that the positional value provides the cell with its position within the system and that this value is used together with the cell's genome to specify its molecular differentiation. This process is defined as the interpretation of the positional information. The set of cells which have their position specified with respect to the same set of points is defined as a field. It can be seen that an essential feature of regulation of fields is the re-establishment of boundary values and the reassignment of positional information [5]. In fact boundaries take on a special importance : what has in the past been referred to as organizer or dominant regions can now be re-interpreted in these terms. In terms of positional information, pattern formation is a two-step process : first the specification of positional information within the field and then the interpretation of this information by the cells which results in the formation of the pattern. With this formulation there is no interaction in a field between parts of the pattern as distinct parts, although cellular interactions and communication are, of course, required for specifying positional information.

The idea that position is important in determining how a cell will differentiate is a very old one, going back at least to Driesch in the 1890s [see 22], and classical gradient theory has long implied [23] that something like position is important. These ideas, however much they may be part of the embryologist's preconscious thoughts, have always been rather vague and have been little analyzed or used in recent years. The concept of positional information attempts to define a much sharper and more useful framework and draws attention to several implications that immediately arise once it is accepted that a cell is having its position specified, and that this information is interpreted in terms of cytodifferentiation.

▶ *Universality and prepatterns.* One of the most interesting features of positional information in relation to pattern is that there is no unique relationship between the form of the gradient in positional value and the pattern which results from the cell's interpretation of it. Using the identical gradient in positional value one can get completely different patterns, simply by changing the rules for interpretation. For example, in a two-dimensional case different rules for interpretation with the same field or coordinate system will give in the one case the French Flag, another the Union Jack, and another the Stars and Stripes (and even the cyto-differentiation is the same in the three cases). There is thus the possibility that the mechanism for specifying positional information might be universal.

This view of pattern formation must be contrasted with those views which explicitly or implicitly claim that in order to make a pattern it is necessary to generate a spatial variation in something which resembles in some way the pattern. For example, the French Flag would require the generation of a step-like function. This view of pattern formation is characterized by the work of Turing [24], Stern [25], Maynard Smith and Sondhi [26], Gmitro and Scriven [27], and Waddington [6, 28], and is the antithesis of the positional information approach. All these workers have considered the appearance of an overt pattern to be the expression of an underlying prepattern. In general, this approach makes use of singularities, that is, local maxima or minima in continuous curves, to specify special features of the pattern. Consider an example based on the work of Stern [25] in which a discrete region is specified along a line of cells—for example, a bristle. In terms of prepattern concepts this requires a local maximum or minimum at this region whereas in terms of positional information this is not necessary. The mechanism for changing the site of the region requires in the pre-pattern case the shift of the peak (whereas in terms of positional information it is the cell's interpretation that alters). This approach logically requires a different prepattern for different patterns and poses severe problems as to how they could be generated.

The concept of a universal mechanism for positional information gets over all these problems and does not require different patterns to be generated by generating a prepattern first. The extreme universalist view, to which I at present adhere, is that the mechanism for specifying positional information and inter-preting it is the same in all multicellular biological systems, not only at the cellular, but at the molecular level. This implies a universal coordinate system. It means that while boundary conditions and sizes of fields may vary, the positional signals and positional values are always the same. Thus all fields would be similar and, from a cell's point of view, indistinguishable, and cells would behave according to position, genome and developmental history. I can imagine that if presumptive amphibian neural tissue were placed at the proximal end of an insect imaginal leg disc it would become forebrain, at the distal end, hindbrain. While extreme experiments of the frog/insect type have not been performed, there is a considerable body of evidence to support the idea of universal fields. The experiments have been done either by grafting or by making genetic mosaics, and in almost every case the cells have behaved according to the predictions of a universal coordinate system : that is, they develop according to their position, genome, and developmental history [5].

The most detailed analysis along these lines has been made by Stern [25] using genetic mosaics. He set out to test if pattern mutants in insects were due to a change in the prepattern or the cell's competence to respond to the prepattern. A theory based on universal positional information must predict that no mutants affecting only one prepattern would be found, since it should not be possible to alter the positional information locally in one field, without affecting all other fields. In every case, except one, he and his coworkers have concluded that the prepattern is unchanged in the mutant, the cells behaving genome autonomously and according to position. The single case where it has been argued that a prepattern change is involved is that of the effects of the mutant eyeless-dominant (ey^D) on Drosophila legs [29], since wild type tissue surrounded by the mutant formed extra structures not found in wild type tissues. A single case demonstrating that a mutant locally alters a field would greatly undermine the concept of a universal coordinate system. It is thus necessary to point out that in the case of ey^D mutants the width of the segment is increased, and the segmentation is disturbed. Both of these could cause an alteration to the boundaries of the field and thus the gradient in positional values, which might produce the observed results [see 3, 30]. It is a situation which requires further detailed investigation.

▶ *Polarity*. Polarity is a rather neglected aspect of development but is central for pattern formation. It is not an easily-defined term in the context of biological forms and patterns, but it is generally used in the sense of implying both an axis and differences along it. It can be regarded like an arrow with a head and tail, or in more mathematical terms it defines vectorial properties. Definitions from the *Oxford English Dictionary* are illuminating : (a) the possession of two points called poles having contrary qualities or tendencies ; (b) tendency to develop in two opposite directions in space, time serial arrangement, and so on. In relation to development I think that the term polarity has been used in two rather different senses. In the first, related to the dictionary definitions, it refers to those situations in which an initially little-structured system becomes increasingly different

88

at what is effectively two ends of an axis. For example, the egg of *Fucus* [31] appears to have no polarity, but one can be induced by an asymmetric environment : this results in the basic polarity of the egg becoming established such that one end becomes the root and the other the plant, the two ends becoming increasingly different. Again in early sea urchin development an inside/outside polarity is established early on and the differences become progressively greater [32]. In the second sense the magnetic analogy is stronger and a vectorial field is implied ; that is, a value and a direction can be assigned at each point in the system, and it is in this sense that it is relevant to positional information. It is a field property and the polarity at each point is defined. This type of polarity can usually be made manifest only by biological tests. For example, the polarity of hydra is defined with respect to which end will form the head and which end will form the foot. It thus refers to the spatial order of the structures which develop. The polarity has a more local and direct visual representation in some systems such as the integument of some insects where the bristles point in a particular direction, or in the amphibian ectoderm where the beat of the cilia reveals the polarity.

There is quite a lot of evidence now available to suggest that field polarity may partly be represented by the sign of the slope of a diffusible substance. This most important suggestion comes from Lawrence [33, 34] and Stumpf [35] from their studies on the insect integument, which suggested that a reversal of the slope of the gradient resulted in a reversal of polarity.

Crick has drawn attention [36] to the confusion that can arise because the word gradient in relation to pattern formation usually means a decreasing spatial concentration of a substance, whereas mathematically it is the slope of the curve defining the concentration at a particular point. As he points out, one can always define a vector in a scalar field : for example, the slope at any point in such a variation in concentration. Thus if we have a decreasing concentration of a substance along a line of cells we at once have a polarity. The converse is not true— one can have polarity without a scalar field. For example, if a line of cells can pump a substance in a particular direction there is a well-defined polarity but there is not necessarily any variation along the line nor is there any difference between the ends. This in fact illustrates two important different ways in which polarity may be established [21]. In the first, polarity is a global property defined by variation in a scalar. In the second, polarity is a local property of the cells, is vectorial, and there is no scalar variation.

There are a number of studies on the early development of polarity in such

89

Positional information and pattern formation

systems as the amphibian nervous system and eye [37], limbs [38], and ear [39] which show that the polarity of the anterior posterior axis becomes determined before that of the dorso ventral axis. Again, in the sea urchin, the animal vegetal polarity is very difficult to alter compared to the dorso ventral one, and the latter appears to be determined later. These time differences are very interesting and raise the question as to whether positional information is specified independently along different axes in a two-dimensional field. The analyses of Cooke and Goodwin [40], and Goodwin [41], suggest that this is not the case.

▶ *Cell movement.* Most current views on the mechanism of directed cell movement or movement of cells to form a pattern is in terms of differential adhesiveness [5, 9, 10, 42]. For example, in our studies of the mechanism whereby the primary mesenchyme cells of the sea urchin embryo moved from their point of entry into the blastocoel to form a relatively well-defined ring, we concluded that it could be accounted for by random pseudopodal activity and a variation in adhesive contact in the substratum over which the cells were moving. The cells would move to those regions where their pseudopods made the most stable contact. This explanation assigns a variation in cellular properties to the substratum which reflects the distribution of the mesenchyme cells, this pattern being effectively a template. In these terms one might expect, and might even look for, a region of increased adhesiveness at the surface of the cells where the mesenchyme form the ring [8, 9]. It is now clear that a different explanation can be offered. Instead of there being a ring of increased adhesiveness in the ectoderm there would be no pattern as such but only the assignation of positional information. Thus moving mesenchyme cells would 'test' the positional information of the cells over which they were moving and would stop at that value for which they were programmed. The pattern would thus again be generated by a cellular response within a coordinate system.

▶ *Mosaic development.* One of the main features of positional information as originally formulated was the idea that intercellular signalling was involved, and is an essential requirement of regulative systems. It is thus essential to examine the so-called mosaic or non-regulative systems within the same conceptual framework, since it seems that here one may well be dealing with a different type of system. The nature of cytoplasmic localization in early development has recently been reviewed by Davidson [18] in relation to the gene activity in early development. While few systems are completely either mosaic or regulative some groups of animals, such as annelids and molluscs, seem to show mosaicism to a high degree. Others, such as the insects, contain both highly mosaic and highly

90

regulative eggs, and it is this latter fact, mosaicism and regulation displayed to varying extents within a single group, that encourages me to continue to look for common features. It is also worth remembering that regulative eggs such as the sea urchin have cytoplasmic localization which may be used to specify boundaries and polarity [3].

From the point of view of pattern formation Davidson's analysis, like the classic work of Wilson [22], is inadequate since both are more concerned with cytoplasmic localization than pattern formation. For pattern formation the apparent absence of intercellular communication is striking. For example, Horstadius [43] showed that after the eight-cell stage the blastomeres of the primitive worm *Cerebratulus* can be isolated or recombined without in any way altering their development. Each fragment develops just as it would have in the intact embryo. The problem of mosaic eggs might be stated as follows : to what extent does the specification of the spatial pattern of differentiation depend on the cytoplasmic differences within the egg and not on intercellular communication, and what is the mechanism whereby cytoplasmic differences are used ? A mechanism based on cytoplasmic differences raises many problems. For example, how accurately must the cytoplasmic localization and the planes of cleavage be specified. This will depend on whether the cytoplasmic differences are graded quantitatively or are qualitatively discrete. The latter would probably make specification much easier since a cell would respond according to the presence or absence of a discrete set of substances. If localization were graded, local intercellular communication at the time of, or just following, cleavage could amplify differences between the two cells. Might one think of a universal cytoplasmic localization system in eggs. One should certainly be hesitant to think of the localization of specific organ-forming substances.

What, one wonders, are the relative merits of signalled positional information and cytoplasmic mosaicism for pattern formation, and why should some eggs be highly mosaic and others highly regulative. Is it partly a matter of distance over which cell-to-cell communication occurs ? It is also curious that some mosaic eggs develop into animals with remarkable regenerative powers.

▶ *Interpretation*. The interpretation of positional information is the key process in the formation of the pattern and is the *raison d'être* for positional fields. It is by this process that the pattern is made manifest. It is not possible to overemphasize its importance since the concept of universality of positional fields places the main burden for pattern formation on the interpretative process. Unfortunately very little if anything is known about it, and this is largely due to our ignorance of the molecular bases of positional value, but interpretation will possibly involve specific gene activation along the lines currently being considered for cyto-differentiation [44]. Certainly advances in the molecular basis of differentiation will greatly help our understanding of interpretation. It is worth remembering that in any positional field the interpretative decisions a cell may take may be relatively simple in the sense that the decision is usually between only a few possibilities. For example, in the developing chick limb the mesenchyme's decision is mainly between cartilage and muscle.

It is also becoming clear that there is a close relationship between determination, as classically defined, and interpretation of positional values. In some cases determination implies that the interpretation will be different : for example, one can think of the determination of wing and leg in the chick in terms of difference in interpretation to a similar positional field [3]. This difference may have arisen in another field at an earlier stage of development. More evidence that positional information may closely relate to a determined state comes from the work of Kieny *et al.* [45] who show that the anterior–posterior level of the mesoderm with respect to vertebral structures and feathers is determined very early on and affects the later interpretation.

One might also view transdetermination [46] in insects in similar terms ; that is, transdetermination involves not a change in the field but in the cells' interpretation of it. A similar view may be taken of homeotic mutants which result in very similar conversions [47].

▶ *Conclusions.* I have tried to show that within the conceptual framework of positional information a new and simple way of looking at pattern formation may be obtained. In pressing the possibility of universality I am deliberately taking an extreme stand, but at least it serves to counterbalance the special-substance-inductive-view of pattern formation. Also in order to show its possible relevance to pattern formation, and even cell movement, I often take a somewhat Pro-crustean view of the data. One of the virtues of the positional information mechanism of pattern formation is that with the same system for positional information one can generate an enormous number of different patterns by changing the cell's rules for interpretation. Since interpretation will be gene determined there is little difficulty in seeing how this can be achieved. In fact, the concept of positional information makes excellent use of a central feature of development, that all cells carry the same genetic information. A further corollary is that it is as easy to make an apparently complex asymmetrical pattern as it is to make a simple one. Pattee has pointed out [48] the necessary complexity for

92

L.Wolpert

making the French Flag using positional information considering that the French Flag is so 'simple' a pattern. This is, however, to miss the point, since while the mechanism is not very simple, the simplicity lies in the fact that it can be used equally easily for apparently complex patterns and the same mechanism may be used for an enormous variety of patterns.

This work is supported by the Nuffield Foundation.

References

1. R.M.Gaze, *The Formation of Nerve Connections* (Academic Press : London 1970).

2. J.Lederberg, *Curr. Top. in Devel. Biol. 1* (1967) ix.

3. L.Wolpert, *J. theoret. Biol. 25* (1969) 1.

4. L.Wolpert, in (C.H.Waddington, ed.) *Towards a Theoretical Biology 3: Drafts* p. 198 (Edinburgh University Press 1970).

5. L.Wolpert, *Curr. Top. in Devel. Biol. 6* (1971) 183.

6. C.H.Waddington, *New Patterns in Genetics and Development* (Columbia University Press : New York 1962).

7. H.Holtzer, in (H.A.Radylcula, ed.) *Control Mechanisms in the Expression of Cellular Phenotypes* pp. 69-88 (Academic Press : New York 1971).

8. T.Gustafson and L.Wolpert, *Int. Rev. Cytol. 15* (1963) 139.

9. T.Gustafson and L.Wolpert, *Biol. Rev. 42* (1967) 442.

10. J.P.Trinkaus, *Cells into Organs* (Prentice Hall : New Jersey 1969).

11. T.E.Schroeder, *J. Embryol. exp. Morph. 23* (1970) 427.

12. N.K.Wessells, B.Spooner, J.Ash, M. Bradley, M.Luduena, E.Taylor, J.Wrenn, and K.Yamada, *Science 171* (1971) 135.

13. L.Wolpert, *Sci. Basis med. ann. Rev. 29* (1971) 81.

14. R.Dulbecco, *Proc. natn. Acad. Sci. 67* (1970) 1214.

15. M.J.Apter and L.Wolpert, *J. theoret. Biol. 8* (1965) 244.

16. K.Sander, *Wil. Roux. Arch. Entw. 167* (1971) 336.

17. P.Weiss, *Principles of Development* (Holt : New York 1939).

18. E.H.Davidson, *Gene Activity in Early Development* (Academic Press : London and New York 1968).

19. H.Driesch, *The Science and Philosophy of the Organism* (Black : London 1908).

20. S.Spiegelman, *Quart. Rev. Biol. 20* (1945) 121.

21. L.Wolpert, in (C.H.Waddington, ed.) *Towards a Theoretical Biology 1: Prolegomena* pp. 125-33 (Edinburgh University Press 1968).

22. E.B.Wilson, *The Cell in Development and Heredity* (Macmillan : London 1925).

23. C.M.Child, in *Patterns and Problems of Development* (Chicago University Press 1941).

24. A.M.Turing, *Phil. Trans. roy. Soc. B 237* (1952) 37.

25. C.Stern, *Genetic Mosaics*, and other essays (Harvard University Press : Cambridge, Mass. 1968).

26. J.Maynard Smith and K.C.Sondhi, *J. Embryol. exp. Morph. 9* (1961) 661-72.

27. J.I.Gmitro and L.S.Scriven, in (K.B. Warren, ed.) *Intracellular Transport* pp. 221-55 (Academic Press : New York 1965).

28. C.H.Waddington, in (S.J.Counce, ed.)

Positional information and pattern formation

Developmental Systems—Insect (Academic Press : New York 1971) ; *and* Form and information, in (C. H. Waddington, ed.) *Towards a Theoretical Biology 4: Essays* pp. 109-40 (Edinburgh University Press 1972).

29. C. Stern and C. Tokunaga, *Proc. natn. Acad. Sci.* 57 (1967) 658.

30. P. A. Lawrence, *Adv. Insect Physiol. 7* (1970) 197.

31. L. F. Jaffe, *Adv. Morphogen. 7* (1968) 295.

32. L. Wolpert and E. H. Mercer, *Exp. Cell Res. 30* (1963) 280.

33. P. A. Lawrence, *J. exp. Biol. 44* (1966) 607.

34. P. A. Lawrence, in *Control Mechanisms of Growth and Differentiation Symp. Soc. Exptl Biol. 25* (Cambridge University Press 1971).

35. H. Stumpf, *Arch. Entw. Organ. 158* (1967) 315.

36. F. H. C. Crick, in *Control Mechanisms of Growth and Differentiation Symp. Soc. Exptl Biol. 25* p. 429 (Cambridge University Press 1971).

37. M. Jacobson, *Developmental Neurobiology* (Holt : New York 1970).

38. R. G. Harrison, *Proc. natn. Acad. Sci. 2* (1936) 238.

39. C. L. Yntema, in (R. H. Williar, P. A. Weiss and V. Hamburger, eds.) *Analysis of Development* p. 415 (Saunders : Philadelphia 1955).

40. J. Cooke and B. C. Goodwin, in *Lectures on Mathematics in the Life Sciences, vol. 3* p. 35 (Amer. Math. Soc. : Providence, Rhode Island 1971).

41. B. C. Goodwin, in *Lectures on Mathematics in the Life Sciences, vol. 3* p. 71 (Amer. Math. Soc. : Providence, Rhode Island 1971).

42. M. S. Steinberg, *J. exp. Zool. 173* (1970) 395.

43. S. Horstadius, *Biol. Bull. 73* (1937) 317.

44. R. J. Britten and E. H. Davidson, *Science 165* (1969) 349.

45. M. Kieny, A. Mauger, and P. Sengel, *Devel. Biol. 28* (In press).

46. R. Hadorn, in (M. Locke, ed.) *Major Problems in Developmental Biology* pp. 85-104 (Academic Press : New York 1966).

47. J. W. Fristrom, *Ann. Rev. Genetics 4* (1970) 325.

48. H. H. Pattee, in (A. Lang, ed.) *Communication in Development* pp. 1-16 (Academic Press : New York and London 1969).

Pattern formation in fibroblast cultures, an inherently precise morphogenetic process

Tom Elsdale

Western General Hospital, Edinburgh

A simple cellular form-building enterprise is provided by a sparse culture of fibroblasts in which the arrangement and distribution of the cells is initially random. As the culture grows the cells interact to form a variety of patterns. By preventing the accumulation of collagen made by the cells, the variety of patterns made by the cells is restricted. It is possible to understand all aspects of pattern generation in these cultures in terms of random movements of cells and rules governing the manner in which cells constrain one another's movements by bodily contact. Pattern generation based on random movements and mutual constraints is a stochastic process. It is suggested that the construction methods used by the cells are essentially the same as the inherently precise methods occasionally used in engineering practice. There is an important restriction on the forms that can be generated by these methods. In addition these methods appear to provide an appropriate language for discussing a number of basic features of development.

▶ *Human fibroblast cultures*. The fibroblast strains are derived from human foetal connective tissue. Cells are grown in Eagle's Medium plus serum in plastic dishes. The cells spend most of their time attached to the plastic and stretched out. In this state they are many times longer than broad. Fibroblasts are motile cells, and move around by a sort of creeping motion that is not understood. The cells can be detached from the plastic and caused to round up into spheres by trypsin. A suspension of these spheres is used to initiate a new culture ; the cells fall out of the medium onto the plastic and they quickly attach and stretch out again. Subculturing in this way is a scrambling procedure, for any order or pattern in the arrangement of the cells before trypsinization is destroyed, and the cells settle out in the new culture randomly distributed and oriented. I have been interested to observe whether this randomness persists as a subculture is grown on and the cells proliferate [1, 2].

Within a day or two one observes that most cells are lying parallel to their immediate neighbours, and when the cells have grown to form a continuous monolayer cultures pass through a 'patchwork' stage where the cells are organized into groups of several hundreds of parallel-arranged cells, the groups being

95

separated from those adjacent in which cells have a different orientation, by narrow packing interstices (*see* Figure 1).

Evolution of the cultures after this stage may take one of two paths depending on whether the collagen synthesized by the cells is allowed to accumulate. The simpler situation occurs in the absence of collagen and this will therefore be described first.

▶ *Pattern generation in the absence of collagen.* Collagen accumulation can be prevented by supplementing the medium with small amounts (c. 60 μg/ml) of bacterial collagenase. This supplement does not affect cell growth, or the attachment of the cells to plastic. A confluent fibroblast culture in a 50 mm dish exhibiting a patchwork pattern contains around $1 \cdot 5 - 2 \cdot 0 \times 10^6$ cells. Growth continues in these cultures until a static population of around 10×10^6 cells is achieved. In the presence of collagenase, this increase in cells is entirely accommodated within the patchwork pattern. The groups of parallel-arranged cells become dense and crowded, and the packing interstices between the groups stand out as empty ditches between raised banks of cells (*see* Figure 2). Actually, the ditches are not truly empty : they contain the clear, granule-free, ruffling membranes of the cells at the peripheries of two opposed groups. The confrontation here is non-parallel and the cells are contact inhibited. We have called the interstices 'frontiers' to indicate this aspect and draw attention to the fact that being contact-inhibited the cells cannot cross a frontier to join an adjacent group [3]. The frontiers are regions of cellular immobility and contrast in time-lapse films with other regions in a culture where the cells are arranged strictly in parallel. Here the cells are free to move, no matter how dense, up and down the array in ways that do not destroy the parallel arrangement. If we regard motility as a scalar term to which a number can be assigned indicating the amount of displacement activities, then frontiers have a low number, parallel arrays a high number.

The divided patchwork characteristic of stationary cultures in the presence of collagenase is not a stable pattern and it undergoes a slow reorganization with time. The cells jostle one another, and the number of frontiers is gradually reduced, and adjacent groups merge as their cells approximate to the same orientation. The end of this process is the transformation of the patchwork into a single uniform parallel array embracing the whole culture and the concomitant disappearance of all frontiers. The global parallel array is entirely stable. The net result of the process is to produce a parallel array from randomness. The process may take months to complete.

A gross explanation of this evolution can be provided on the basis of three factors.

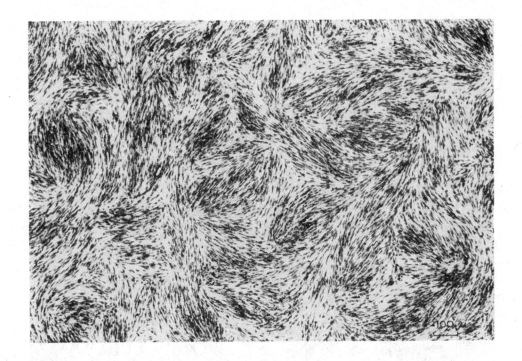

FIGURE 1

Confluent monolayer culture of human lung fibroblasts. Three days after random initiation with about a third the number of cells here present. The arrangement of the cells is no longer random. The cells form a rough patchwork of local parallel arrays termed groups. It is the groups that lie at random rather than the cells

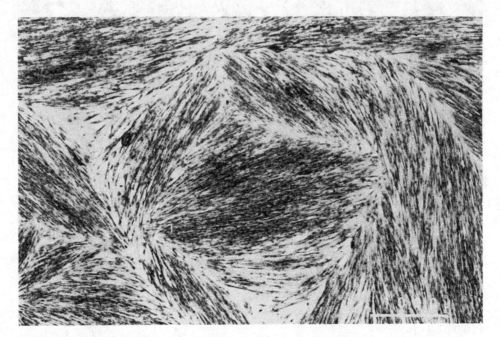

FIGURE 2

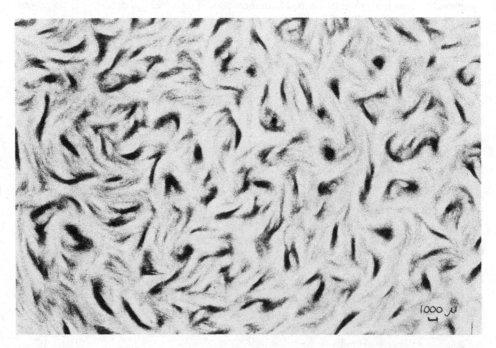

FIGURE 3

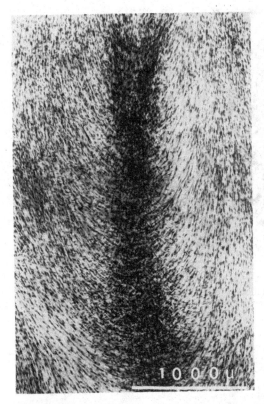

FIGURE 4

Stationary culture, human lung fibroblasts. Higher power picture of a ridge from Figure 3.
Ridges are initiated as primary overlaps forming near or over frontiers in post-confluent cultures.
Cells are recruited into the aggregations by migration from adjacent underlying groups. The
direction of movement of these cells is indicated by the U shaped curves, by which the two
ends of a ridge can be distinguished as 'convex', and 'concave'. The ridge is bilaterally symmetrical. There is an upper surface protruding into the medium, and a lower surface on the
plastic ; left and right sides can be assigned by convention. This ground plan is the same as
that of the vertebrate body

FIGURE 2 *(facing)*

Human lung fibroblasts grown to their stationary density in the presence of collagenase.
Patchwork pattern accommodating all the cells produced by post-confluent growth. The
groups are dense and crowded, the cells demonstrating good parallel alignment. The boundaries
between groups stand out more clearly than in Figure 1. They are termed frontiers to indicate
that they are regions of cellular immobility and contact inhibition. Frontiers form where ranks of
cells make a non-parallel confrontation. Cells cannot cross a frontier to join those in another
group—not in the presence of collagenase

FIGURE 3 *(facing)*

Human lung fibroblasts grown to their stationary density without collagenase. This low power
picture demonstrates a markedly uneven distribution of the cells, quite different from the
situation shown in Figure 2. There are numerous, mainly ridge-shaped aggregations formed as a
result of directed cell migrations

FIGURE 5

Stationary culture of human lung fibroblasts. A high power photograph of a portion of the sloping side of a ridge to show the characteristic internal pattern. Ridges comprise stacks of orthogonally arranged monolayers of cells (*see* text)

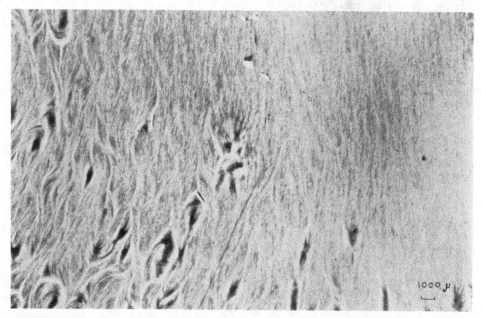

FIGURE 6

Stationary culture of human lung fibroblasts maintained for several weeks and showing disappearance of ridges. Over about half the area of the photograph the cells have dispersed from the ridges and moved into the lower layer of the culture to form a uniform parallel array oriented along the short axis of the picture

1. The cells are elongated—necessary if a parallel array is to be detected.
2. The cells are motile. It is going a little beyond this basic observation to suggest that the cells are inherently motile, implying by this that cells are customarily in motion except when prevented from moving. It is also assumed, in the absence of evidence to the contrary, that cell movement is randomly directed except in so far as movement in particular directions is constrained by the presence of other cells and the side of the dish. This assumption appears valid in the case of 'Falcon' (Becton Dickinson, Brindley 74, Astmoor Industrial Estate, Runcorn, Cheshire) plastic dishes.
3. There is a rule governing the way the cells constrain one another's motions by bodily contact. Cells cannot move over and across one another. Scalar motility is strongly constrained by non-parallel confrontation, less constrained by parallel confrontations.

On the basis of these factors, the cells in a dense fibroblast culture can be envisaged as continuously exploring their immediate environments for permitted avenues of movement. The result of this activity, over a long period of time, will be that the cells will come to adopt that configuration which allows them the maximum exercise of their motility. This configuration is the parallel array without frontiers.

▶ *Pattern evolution in the presence of collagen accumulation.* Whereas post-confluent growth in the absence of collagen was accommodated in a two-dimensional pattern, in the presence of collagen this growth engenders three-dimensional structures (*see* Figure 3), in the following manner.

Almost as soon as confluence is attained, multilayering commences. Multi-layering occurs where cells are observed lying above and across others. Note the terminology here, a parallel array may be several cells thick but it is not multi-layered because all the cells share the same orientation. Multilayering is first observed in the form of primary overlaps either over or around frontiers, they are never observed away from frontiers. These overlaps develop into thick ridges with bilateral symmetry (*see* Figure 4). The ridges are formed entirely by cell migration. Differential growth is unimportant, for if collagenase is removed from stationary, collagenase-grown cultures, ridges form rapidly in the absence of net growth. The characteristic internal pattern of the ridges is the orthogonal array. This consists of a multilayered stack, each layer is a parallel array, arranged perpendicularly to the layer beneath and above (*see* Figure 5). In the vertical axis the pattern repeats every two layers starting at any point. These formations are metastable, they are not maintained for long periods. The cultures undergo

97

essentially the same changes as those in collagenase (see Figure 6), and eventually provide global parallel arrays of uniform thickness.

Multilayering clearly requires collagen, and there is good evidence that the collagen is located as an intervening band between each layer of cells. However, the amount of collagen is sufficient to cover only about 1/50th of the surface of the cells, so it is not appropriate to think of the collagen as providing an insulating material effectively isolating one layer of cells from contact with those above and beneath it.

Turning now to mechanisms, I shall ignore gross aspects of pattern, including the symmetry and polarity of the ridges, and focus attention on the heart of the matter—the genesis of orthogonal multilayers and the role played by the collagen. Clearly the collagen plays a permissive role here, for orthogonal multilayers do not form in the presence of collagenase. In the latter case cells are prevented from moving over one another by contact inhibition, in the absence of collagenase such movements become possible. However it is well known that fibroblasts align on collagen substrates, and it is possible that collagen serves a second function in the genesis of orthogonal multilayers dictating the orientation of the cells.

Figures 7a and 7b illustrate two ways in which orthogonal multilayers may be formed. Both assume a permissive role for the collagen. Consider Figure 7a first. Here the cells in a localized region of a monolayer are pictured secreting a collagen substrate above themselves. This substrate is available for colonization by cells from beyond its boundary. It is assumed that the collagen fibres are laid down with a particular orientation causing the colonizing cells to align in a particular manner upon them. The cells in this second layer, once in place, themselves secrete a collagen substrate above themselves, the orientation of the fibres bearing a fixed relation to the orientation of the cells. In this way a stack can be built up. This scheme envisages cells aligned by collagen and collagen by cells. What rules are necessary if the result is to be an orthogonal multilayer? There are two possibilities: cells secrete collagen fibres parallel to themselves, colonizing cells orient preferentially, perpendicular to the fibres; alternatively, the reverse situation: collagen fibres are extruded perpendicularly to the orientation of secreting cells, and colonizing cells align parallel to the fibres. This results in a paradox, for the first possibility can be disproved by a simple experiment, and the second is incompatible with the structure of muscle and tendon. Furthermore no explanation is offered for the failure of orthogonal multilayers to arise in parallel arrays, nor is the initiating role of frontiers in the generation of primary overlaps accounted for.

98

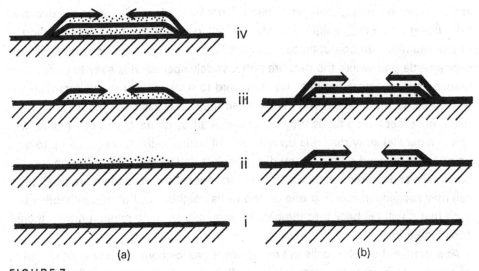

FIGURE 7

Diagrammatic vertical sections through a fibroblast culture to illustrate two hypotheses to account for multilayering. Both hypotheses provide for multilayers structured by alternate layers of cells and collagen. Both also assumed a permissive role for collagen, allowing cells to move over others without hindrance from contact inhibition. Solid lines represent monolayer sheets of cells ; dots, collagen. (a) This hypothesis allows for a directing role for collagen in the formation of orthogonal multilayers, this aspect is not illustrated in this diagram, but *see* text and Figure 8. (i) Monolayer of cells. (ii) Collagen secreted above an area of the monolayer. (iii) The collagen substrate so provided is colonized by cells from outwith its boundaries. (iv) A third layer forming by the same mechanism. This hypothesis leads for paradoxes and is untenable, *see* text. (b) This hypothesis allows for no directing role for collagen in the final arrangement of the cells in the multilayers. (i) Monolayer of cells. (ii) Cells secreting collagen beneath themselves and able to move over other cells in any direction except parallel to the cells beneath them ; in the latter case the overriding cells 'fall' into the parallel array and become incorporated in the layer beneath. (iii) A third layer forming by the same mechanism.

Figure 7*b* illustrates another way orthogonal multilayers might form. Here it is assumed that cells extrude collagen, not above themselves as in the previous hypothesis, but beneath themselves. Cells grow 'collagen caterpillar tracks' ; furnished with these, cells are enabled to move over and across others. This is to grant collagen a permissive role in multilayering, but although cells may well leave a trail of aligned collagen fibres behind them, the direction of movement of the cells and their orientation is in no way dictated by collagen. This scheme therefore differs from the last in assigning no directing role to the collagen. Under this hypothesis orthogonality has to be accounted for directly in terms of random movements constrained by cellular interactions.

Consider first the stability of parallel arrays, bearing in mind that orthogonal multilayers are never spontaneously generated within these formations. The cells are assumed to possess collagen caterpillar tracks ; they are therefore not debarred from riding up on their neighbours and migrating over them. My understanding of this situation can be put across with the help of an analogy. Consider walking over a cattle grid where the bars are rather widely spaced. It is easy to walk perpendicularly across the bars, but awkward to walk parallel to the bars with the likelihood that the feet will fall between them. It is assumed that cells move randomly except in so far as they are constrained by contact with other cells. A cell in a parallel array may ride up onto neighbouring cells. It is flexible up to a point and may bend a little in any direction, but its chances of moving on top of other cells and bending sufficiently to lie orthogonally across them are small ; the cell may succeed in crossing one or two of its neighbours, but the probability is high that it will fall back into the parallel array and become reincorporated therein (*see* Figure 8*a*).

At a frontier, however, cells in two adjacent groups confront one another out of parallel at an angle that may vary from 90° to a minimum around 20°. The greater the angular difference between the orientations of the cells in adjacent groups, the greater the likelihood that cells moving with their initial orientation across the frontier and bending somewhat to either side will create an orthogonal overlap (*see* Figure 8*b*). Non-orthogonal as well as strictly orthogonal overlaps will be formed in this way, but the former will be less stable than the latter. This follows because less movement will be required to bring the cells in a non-orthogonal overlap parallel to the cells in the layer beneath, causing them to be lost to the overlap by incorporation into the underlying layer. Selection against non-orthogonal overlaps is borne out by observation, for although non-orthogonal overlaps are occasionally observed, most of the overlaps seem approximate to orthogonal. However even strictly orthogonal overlaps should be unstable and decrease in extent with time as peripheral cells fan out at random (*see* Figure 8*c*). This too is precisely what is observed, for orthogonal multilayers are metastable structures ; they arise in post-confluent cultures growing towards their stationary density. They persist for a while, and gradually disappear by the

---

FIGURE 8
These surface view diagrams illustrate the three features that characterize multilayering in fibroblast cultures. (a) *Stability of the parallel array,* (aii) Cells moving out of the array have a high probability of 'falling back' and becoming reincorporated. (b) *Overlapping across frontiers,* Orthogonal overlaps form naturally across frontiers between two groups of cells whose orientations differ by 90 per cent [Non-orthogonal overlaps form in situations intermediate between (a) and (b).] (c) *Metastability of multilayers,* Fanning out of an orthogonal overlap bringing cells parallel to those beneath, thus causing cells to be lost to the overlap according to (a) above. Non-orthogonal overlaps are less stable than orthogonal overlaps.

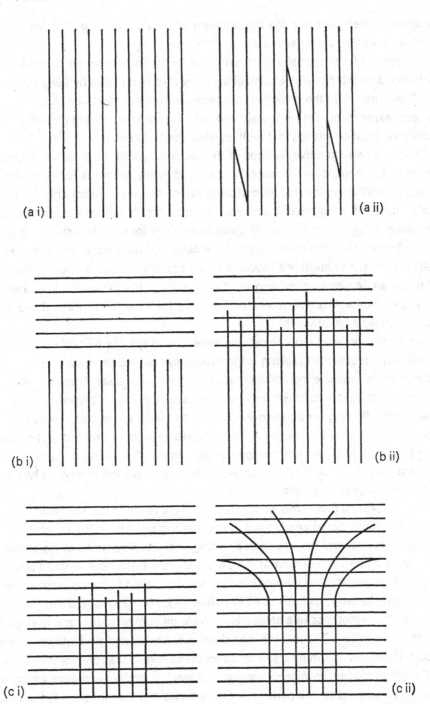

dispersal of their cells into the bottom layer of the culture, as part of the evolution towards a single stable parallel array.

The phase of orthogonal multilayering can be considered as an interlude in the evolution towards the single parallel array, evoked in cultures by fairly rapid post-confluent growth in the presence of numerous frontiers. It can be shown by suitable experiments that in the absence of either post-confluent growth or numerous frontiers orthogonal multilayering does not occur.

There are two attractive features of this second hypothesis to account for the genesis of orthogonal multilayers. First, it incidentally accounts for the stability of the parallel array, the initiation of overlapping around frontiers only, and the metastability of orthogonal structures. Secondly, this hypothesis is formulated in the same terms as that previously given to account for parallel arrays, that is to say in terms of random movements of the cells, and rules governing the way the cells constrain one another's movements via contact. Indeed a single general set of rules can be written providing for the situation in the presence of collagenase as a special case. It is not necessary, however, for what follows that these rules be formulated explicitly here.

A scheme involving random cell movements and specific mutual constraints is a stochastic process. In seeking to generalize further and produce a conceptual framework for considering cellular pattern-forming processes, there are two requirements. The facts as known must be respected, hence analogies with matchsticks floating on troubled waters or iron filings in magnetic fields are un-acceptable because we know that cells are self-energized, and matchsticks and iron filings are not. Second, the conceptual framework should be appropriate in the first place to the biological context, it does not necessarily have to be funda-mental in terms of physics or physical chemistry.

The conceptual framework suggested here is in terms of the so-called inherently precise methods occasionally used in engineering practice in the construction of exceedingly accurate devices such as ruling engines. Rowlands, in his article on the screw in the 1913 edition of the *Encyclopaedia Britannica*, emphasized the primitive nature of these methods, and gave his opinion that they were indeed the pattern-forming methods used in nature [4]. More recently, John Platt has invoked these methods in a stimulating article on vernier acuity, and the develop-ment of vision [5]. The ideas involved are very simple and well known to every-body. The reason for referring to these methods here is that they help in forming a Gestalt of pattern-forming processes that employ random energetic inputs and mutual constraints by the parts [6]. Furthermore there are inherent limitations on what can be achieved by such systems. These limitations are part of the concept and Platt gives a more detailed approach to this aspect in terms of a functional geometry derived from group theory.

▶ *Introducing the inherently precise method.* Consider the operation of shaping things. A typical device that man uses for this purpose is the lathe. The accuracy of the work is limited by the tolerances to which the lathe itself is constructed. Ideally all uncertainties in the control and operation of the machine are banished.

There is however another way of shaping things. Suppose you wish to make a lens with a perfectly spherical surface, and incidentally you are not too fussy about its final size : take two rough blanks of glass, one convex, the other concave, of roughly the same radius of curvature. Fit the convex into the concave, and rig up some way by which the two faces can be rubbed together. If this is done in the right way then the blanks bed ever more perfectly into one another. Spherical surfaces will be generated to any desired degree of accuracy depending on secondary factors such as the fineness of the grinding paste used and the patience of the overseer. It all really depends on rubbing the blanks together in the right way. The condition that has to be met is that grinding motions of one type, say a North–South motion of one of the blanks on the other, do not outnumber other motions. This is provided for if the grinding motions are random. This then is a machine for generating spherical surfaces, and it works in a quite different way to the lathe.

We have noted the random energetic input, essential in the lens-grinding machine and vitiating in a lathe. Indeed to secure them in the lens-grinding machine one needs loose, flexible joints with plenty of play, quite different to the accurately machined parts of a lathe. Furthermore the lens-grinding machine is potentially more accurate, as accurate as you want it to be ; the method is inherently precise. Inherently precise machines cannot be made to generate any shape ; only those shapes that show displacement congruence can be made by these methods. Note that two complementary spherical surfaces of the same radius can be apposed so that they are in contact over all overlapping points—congruence ; and remain so as the joint so formed is moved—displacement congruence. In an inherently precise machine, pattern is perfected gradually over the surface of the blanks ; the lathe on the other hand is a scanning machine.

▶ *The generation of parallel arrays of fibroblasts as an inherently precise process.* Consider now the randomly initiated fibroblast culture as a machine for generating parallel arrays of fibroblasts. It is simpler, to start with, to think of the cells growing in the presence of collagenase.

Pattern formation in fibroblast cultures

Note first that the energetic input is random. This is provided for by the individual and independent energizing of the cells. It is assumed also that cells are not biased in their movements except in so far as they are constrained by other cells.

Second, there are durable rules governing the mutual contact constraints on the motions of the cells. Similar rules govern the way the parts interact in the engineers' machines. There the rules are defined by the way the parts are articulated, and vary from one operation to another—making spheres, flats, screws, cogs, and so on.

Third, there is no intrinsic limitation in the method on the accuracy to which parallel arrays of fibroblasts can be generated.

Fourth, the pattern is perfected gradually over the whole dish.

Fifth, note that the parallel array is a dynamic pattern, the cells move up and down the array reaffirming the pattern by their movements. The parallel array exhibits displacement congruence in life. Here is the inherently precise method displayed in a cellular pattern-forming context.

Clearly fibroblast cultures are very different from lens-grinding machines, and a little thought is required to subsume both under the principles of the '*IP*' method. Note that the lens-grinding machine will work given only very small grinding movements about some arbitrary starting position of the joint formed by the apposition of the two blanks ; large traverses are not required. This is analogous to the situation in a cell culture, where the individual cells do not range over the whole dish but probably stay within the square millimetre or so about their initial locations. Note also that orthogonal multilayers although generated by a random energetic input and specific constraints do not form part of an inherently precise process, because displacement congruence is not defined on them. Continued cell movement under the same constraints as generated orthogonal multilayers leads to their eventual unravelling.

The inherently precise method appears appropriate for the consideration of local pattern-forming processes involving one or several cell types as exemplified in the generation of local tissue architecture *in vivo* and sorting out experiments *in vitro*. The *IP* method formulates in a general way the minimal requirements necessary in order to envisage the automatic generation of stable pattern in these systems, occurring without outside direction. It was probably this aspect of the methods that impressed Rowlands and suggested to him adjectives like simple, primitive, and natural to describe these methods.

It is interesting to consider in a general way certain very obvious and basic facts about development in the light of *IP* methods, not because explanations are thereby provided, but rather because the methods offer a convenient language for such consideration. In what follows, several general points are touched on.

▶ *Wearing in and out*. Man-made machines, such as automobiles, wear out in use and are conserved in disuse under ideal storage conditions. It is a general feature of living things on the other hand that they reaffirm their functional capacities in use ; unused these capacities wither. A topical example of this statement might be provided by the deleterious effects of prolonged weightlessness on astronauts. The statement also has validity on the longer time scale of evolution. Turning now to fibroblast cultures one notes that the stable parallel array is a dynamic pattern reaffirmed by the continued movement of the individual cells within it. This movement is the eraser that smooths away blemishes and keeps the pattern fresh. It is easy to make small wounds and mutilations in a parallel array and observe their rapid effacement. If the cells did not move, reaffirmation of the pattern would not occur, and thermal agitation would eventually disrupt the pattern. This situation is analogous to the conservation of a functional capacity in use and its withering in disuse. It is characteristic of *IP* processes that they possess some property analogous to conservation by use.

▶ *The internal access machine*. Man's machines are designed for external access, for external construction and control. The record of the operation of such a machine is usually written outside the machine itself. The record of the operation of a lathe is the finished piece of work taken from it. Ignoring wear, the machine is not changed in doing the work.

On the other hand, the hen's egg develops autonomously. The egg, we might say, is a machine designed for internal access, for mutual access (interaction) between the parts. The record of the operation of this machine is written internally, in directional changes within itself—development. The machine is changed by its operation. The same is true of any inherently precise machine.

▶ *The world of small things*. Cells live in a less stable world than man. Objects which are large enough to appear significant in the small world of the cell are yet small enough to display Brownian motion—bacteria and mitochondria for example. The cell appears to manage in an unstable world by relying on certain highly stable features of its constitution and muddling through for the rest. The more directly a property is under genetic control the more that property will partake of the high stability of the genetic information. States of determination and differentiation are, as a rule, stable. Features less directly under genetic control will be subject to uncertainties. The following simple and easily-repeated experiment suggests uncertainties in the times cells may take to accomplish things.

Plate a few hundred normal fibroblasts into a petri dish and ten days later some of the cells will have proliferated into colonies. Some colonies will be big, others small, the variation in colony size being great. Pick the largest colony and reclone ; you will grow some large and some small colonies. The variation will not have been manifestly reduced : you have not selected for faster growth and a higher cloning efficiency. Precise quantitative observations along these lines have led Merz and Ross to the conclusion that the growth of normal fibroblasts in culture obeys the stochastic simple birth process [7]. The statistics are formally analogous to those used to describe radioactive decay in an isotope. A population of normal fibroblasts under controlled conditions will display a characteristic and predictable doubling time analogous to the half life of an isotope. The stochastic simple birth process however implies that the growth of any one cell is no more predictable than the decay of a single atom of isotope. Many of our quantitation measurements refer to average cell behaviour ; the large variation in the behaviour of unit cells is sometimes forgotten.

In addition, uncertainties are to be expected in the way cells move, especially in the absence of stable isotropies in the environment.

Instabilities due to small size are not confined to the small world of the cell, for they show through into the macroscopic world of living plants and animals. The physicist may be able to repeat his observations to an accuracy of a millionth part of 1 per cent, by virtue of his ability to reduce uncertainties in the operation of his measuring devices to very low levels. The biologist, on the other hand, is often well satisfied if he can repeat his experiments to an accuracy of 10 per cent. No two oak leaves are identical. It appears therefore that the gradual emergence of macroscopic structure in morphogenesis is not accompanied by a progressive emancipation from the instabilities inseparable from the world of small objects.

To be sure, the macro structures produced by morphogenesis are stabilized in various ways, in particular by the deposition of bone and other matrices in the case of animal development. Such stabilization is seldom carried to the lengths of vitiating the recovery of a measure of plasticity in the event of injury. However turning to the beginnings of morphogenesis, such stabilizations are inappropriate, freedom of movement is the order of the day. We conclude from the previous arguments that it is not a requirement for morphogenesis that the behaviour of cells be first stabilized in all particulars ; only the determination and pathway of differentiation is stabilized, leaving other aspects of cell behaviour subject to the instabilities inseparable from the world of small objects.

▶ *Morphogenesis*. Morphogenesis is ultimately mechanics—it involves individual cells arranging themselves into patterns. It is precisely in this enterprise that uncertainties in cell behaviour would appear to be vitiating, serving to combat and sabotage the order-creating processes of morphogenesis. This view however is an anthropomorphism based on the recognition that many of the constructional methods used by man, satisfactory in his larger world, would give unsatisfactory results transposed to the world of small objects. We need to look for construction methods appropriate to the world of small objects and available to cells. The appropriate techniques would appear to be the inherently precise methods. These allow us to visualize the uncertainties in cell behaviour being institutionalized for constructive ends as contributions to the random energetic inputs which are a necessary feature of these constructional methods. Thus it makes no difference whether we have identical fibroblasts whose behaviour is in all respects predictable, or fibroblasts exhibiting large chance fluctuations in behaviour. Parallel arrays will be equally well generated by both given the inherently precise method, granted only that the rules governing the way the cells constrain one another remain constant. The constancy of the rules is insured by the fact that they depend on the specific properties of the cell surface, and these are under direct genetic control. (Note that the rules are usually radically rewritten following the introduction of a little novel nucleic acid in the cells, in the form of a transforming virus.)

▶ *Maintenance and regulation*. Randomly initiated fibroblast cultures evolve to parallel arrays ; the inherently precise method allows us to visualize this evolution occurring automatically so long as the cells continue to move. In addition it is to be noted that mutilation of the final pattern does no violence to the pattern-forming processes. Trypsinize a parallel array and set out the cells at random again, and they immediately embark on generating another parallel array. A fibroblast culture, whatever the configuration of the cells, is inevitably at some point on a trajectory directed towards the parallel array. Systems exhibiting such trajectories are referred to by Waddington as chreods [8, 9]. All developmental processes in life are held to be chreods. The simplest visualization of a chreod is in terms of the basic requirements for an inherently precise process.

On the view that all morphogenetic processes can be considered as inherently precise processes, the function of the genome in morphogenesis is to specify the rules governing mutual cellular constraints in each temporal and spatial

Pattern formation in fibroblast cultures

compartment of development. This, however, is unlikely to be the sole function of the genome in morphogenesis, for something has been left out. Using inherently precise methodology and given a control hypothesis for unlocking the genetic information we could envisage a succession of morphogenetic processes taking place in the embryo and its parts. But these separate processes are coordinated and integrated to comprise the embryo, and a total view is necessary. Justice has not been done to the embryologists' descriptive idea of the embryonic field. The total view will understand the embryo as the entity that does the developing, rather than as the summation of part processes.

To suggest that the whole embryo is inherently precise is to suggest at best a useless platitude, equivalent if true to saying that all the energy we use in our bodies is ultimately derived from the sun—not a very helpful comment to make to a colleague working on mitochondria.

References

1. T. R. Elsdale, *Exp. Cell Res. 51* (1968) 439.
2. T. R. Elsdale, *J. Cell Biol. 41* (1969) 298.
3. M. Abercrombie, *In Vitro, 6* (1970) 128.
4. H. A. Rowland, The screw, in *Encyclopaedia Britannica* (11th Edition 1913).
5. J. R. Platt, in (H. P. Yockey, ed.) *Symposium on Information Theory in Biology* (Pergamon Press : Oxford 1956).
6. T. R. Elsdale, in (G. E. W. Wolstenholme and J. Knight, eds.) *Symposium on Homeostatic Regulators* (J. and A. Churchill : London 1969).
7. G. S. Merz and J. D. Ross, *J. Cell Biol. 35* (1967) 92A.
8. C. H. Waddington, *New Patterns in Genetics and Development* (Columbia University Press : New York 1962).
9. C. H. Waddington, in (C. H. Waddington, ed.) *Towards a Theoretical Biology 1: Prolegomena* (Edinburgh University Press 1968).

Form and Information

C. H. Waddington
University of Edinburgh

One often hears it said that today, with the genetic code cracked and the mechanism of protein synthesis well on the way to being understood, there remain only two major problems in biology : morphogenesis and the operations of the central nervous system. I shall here discuss only the first of these.

To speak of 'the problem of morphogenesis' implies that there is some single crucial secret to be discovered, which, when known, will indicate the lines along which all morphogenetic phenomena are to be interpreted. If this is the situation, then the most important immediate task is to search out, and come to experimental grips with, a suitable paradigm case of morphogenesis in which the essential basic mechanism can be uncovered. Several people, such as Crick, Wolpert, Goodwin and Cohen, are in fact doing just this. They will surely discover many very interesting and important things. However, I am personally rather sceptical that there is such a thing as an essential morphogenetic mechanism, or that any one example can serve as a paradigm for the whole range of instances in which forms or patterns are produced. To my mind, the evidence suggests that there are, at work, several different processes of radically different types.

One reason why the idea of a unified theory of morphogenesis is so attractive is the feeling that the particular properties of life depend very crucially on *form*. If one could discover a 'secret' of form, that would have a good claim to be one of the secrets of life, and a major one at that ; whereas if form arises by quite different processes in different instances, can it really be so fundamental after all ? The answer to this is that a great deal of morphology is not of very fundamental importance. Whether a fly's leg has five segments or four, whether the hairs on a bug's thorax are more, or less, evenly spaced, may perhaps have some moderately important consequences in relation to natural selection, but do not rate as among the basic problems set by the existence of living things. One is tempted to grant them such importance only because of a confusion between some of the numerous contexts in biology in which that very general word 'form' is used.

In all contexts, one can take 'form' to mean a specific, that is, recognizable, three-dimensional configuration of matter. I think we recognize that one of the most important characteristics of the basic mechanisms of biological systems is

Form and information

that they do involve specific three-dimensional forms (which in fact are always parts of four-dimensional, time-extended sequences).

This point is concealed by the conventional description of biological systems in terms of the processing of information. It is a fact that living things incorporate their most essential information in *two*-dimensional, not *three*-dimensional, codes, such as the linear sequences of nucleotides and amino-acids in nucleic acids and proteins. But one cannot conceive of systems of processing this information, of transcribing it, translating it, filtering it, doing to it anything else you can do to information, by mechanisms which remain strictly two-dimensional. One could perhaps imagine something like a Turing machine with a reading device and a means of moving tape backwards or forwards which was confined almost to the plane of the tape, and made little use of the third dimension ; but it would have to make *some* use. But this is not what biological systems have done. The information-containing molecules, as we meet them in the flesh, have a very well-developed 3D form. That nucleotide sequences normally occur as 3D double helices might admittedly be regarded as a trivial matter of chemical convenience ; all the logical needs would be satisfied if they appeared as paired complementary strands lying in one plane. But the fact that the amino-acid sequences of proteins fold up into a 'globular' protein with a tertiary structure is of fundamental significance. It is only thanks to this extra dimension that they can function not merely as repositories of information in which no one is interested but as machines which can interpret the nucleic-acid-borne information as 'instructions' or 'programmes' to operate on and change appropriate parts of their external world in definite ways. It is only because linear sequences of amino-acids fold into tertiary structures that enzyme action is possible ; this 3D structure provides the basic non-holonomic constraints whose importance Howard Pattee has emphasized in earlier meetings.

Nucleic acids, also, can tie themselves up into knots or 'tertiary structures', for instance if they comprise a single strand which contains a number of sequences which can pair with one another, separated by stretches which have no complementary partners ; as is the case with transfer-RNAs. The bonds between the paired sequences would in general be expected to be rather strong, so that the possibilities for allosteric behaviour would probably be much less than in the tertiary structures of proteins. It might be, however, that such 'tertiary nucleic acid structures' are involved in the recognition reactions between nucleic acids and proteins. But there is no evidence that they can act as programme operators ; there are no nucleic acid enzymes.

C. H. Waddington

Now generation of these tertiary structures of proteins does depend on one specific mode of form-generation, namely chemical interactions between atoms or small groups of atoms, leading to the formation of such links as disulphide bridges and stereo-specific non-covalent associations. One could, if one wished, apply the name 'morphogenesis' to these processes, perhaps qualifying it as 'molecular morphogenesis'. And they are certainly fundamental to biology, in the sense that no living system, of the kind we encounter on this earth, could operate if linear information-carrying molecules could not be moulded into specific three-dimensional forms.

I am arguing that only in this sense is morphogenesis really fundamental to biology. Certainly it is an almost—or even completely—universal characteristic of living systems, but in all realms other than the molecular, biological forms seem to me to originate as *consequences* of biological activities, rather than being primary *causes* of them. Of course, since living processes are extended in time, and usually proceed through several steps, a form which arose as a consequence in an earlier stage will act among the causes of later events. It may in this way be of great importance (as for instance the morphogenesis of neural connections is in the operations of the brain, or in the sense in which the harmfulness of cancer can be attributed to abnormal morphogenesis of tissues). But important things are not always fundamental. Electric power in the grid is important, in the sense that modern civilized life could not be carried on without it, but it is not among the fundamental entities of physics, being only a consequence of their mobilization in a particular way.

Morphogenesis at supra-molecular levels seems to me to be an inevitable result of three facts about biological processes. First, such processes go on in a medium—protoplasm, it used to be called—which is liquid enough to allow fairly rapid diffusion and chemical reactions, but viscous enough to offer some resistance to deformation. Secondly, biological reactions are of kinds which may cause appreciable changes in the physical properties of this colloidal medium, including alterations of density and volume. Finally, all biological systems involve some processes with non-linear or even threshold characteristics, which guarantee that, under the influence of stochastic factors, rates at which the processes are proceeding will not be uniformly distributed throughout a mass of living material, even if this were (in an imaginary ideal case) supposed to be homogeneous to start with. These three characteristics are enough to generate non-uniform stresses, to which the living material must respond by assuming some form or other.

111

Form and information

The most basic questions which arise, on this view, are : first to list and classify all the various ways in which the physical properties of the 'protoplasm' may be altered ; and secondly to account for the extreme repeatability of most biological forms. As to the first problem, I made a tentative beginning some years ago [1] when I suggested that there are four main categories of form-generation. These can be arranged as follows :

1. *Unit-generated forms*. The forms are produced by the interaction of certain unit elements, and their character is determined by the properties of the units.

(a) Particle systems : the units can be treated as small volumes. The smallest particles are molecules, as in the 'molecular morphogenesis' discussed above. One can also have larger particles, such as those involved in the morphogenesis of viral structures or bacterial flagellae, ranging right up to a possible self-assembly of groups of cells, which might assort themselves according to the position of specific adhesion sites such as desmosomes, although I am not aware that this has yet been conclusively demonstrated.

(b) Fibre systems : the units are one-dimensional.

(c) Sheet systems : the units are two-dimensional.

2. *Instruction-generated forms*. The structures are produced from a set of units plus a set of instructions as to how they should be assembled.

3. *Template-generated forms*. The structures take their basic form from some possibly simpler existing formed structure.

(a) An exact copying or simple coding exists between template and copy.

(b) Template production of noncopies ; that is, the coding between template and copy is highly complex.

4. *Condition-generated forms*. The structure arises as the working out of an initial spatial distribution of interacting conditions.

(a) Stochastic conditions.

(b) Determinate conditions.

I will not pursue here the reasoning concerning the repeatability of biological forms, beyond remarking that it seems to depend essentially on the pre-existence of a structure in the region in which the form in which we are interested is putting in an appearance, which can either impose geometrically defined boundary conditions or induce certain localized centres of activity.

The main subject discussed at this Symposium has been the arising of forms which would fall into the category of 'condition-generated'. The examples con-sidered were of a kind which can, without too much loss, be reduced to two-

112

dimensional arrangements of discriminable elements, a category which can be referred to as static patterns. The main theoretical outlooks advanced were three.

The most general of these is Wolpert's concept of 'positional information' [2-4]. The idea is intended to be applied to situations in which the development of a cell is dependent on its position within some mass of tissue. Wolpert suggests that it would be easy to understand such behaviour by assuming that the cell has some method of ascertaining its position within the mass (its three co-ordinates on some systematic frame of reference), and is provided with a code which enables it to adjust its performance to accord with this 'positional information', and a clock which tells it when to do so. Several authors [for example, 3, 5-8] have suggested various ways in which positional information might be supplied to a cell. Closely allied to the notion of positional information is the well-known concept of 'gradients' ; in fact, a gradient may be considered as one of the ways in which positional information may be supplied.

Another idea which has been widely discussed in recent years deals with patterns which are just this side of complete disorder, since they consist of entities such as hairs or bristles which are arranged at random, subject to the constraint that there is a minimum separation between two nearest neighbours, or some similar constraint involving all near neighbours [9, 10].

Finally, forms or patterns must be discussed in connection with the notion of buffered developmental pathways. It was in fact from studies on the patterns in Drosophila that the idea was first developed that developmental processes in higher organisms generally involve the synergistic action of many genes, whose operations interact in such a way as to define a pathway of change in a multi-dimensional phase space which has a type of stability, in that after disturbance the developing system tends to find its way back to a later part of the normal pathway. Such stabilized pathways are the 'chreods' which we have often discussed here.

CATEGORIES OF PATTERN ALTERATION

The categories to be distinguished are of quite a general character, but they will be more easily envisaged with some definite examples in mind. We shall use the pattern of a four-of-diamonds playing card as the basic form whose alterations are to be distinguished (Figure 1).

In the first place, it must be noted that such a pattern consists not only of the *pattern elements* which are arranged in a certain way, but also of the *pattern region*, that is, the area of the card. A similar distinction needs to be made in

Form and information

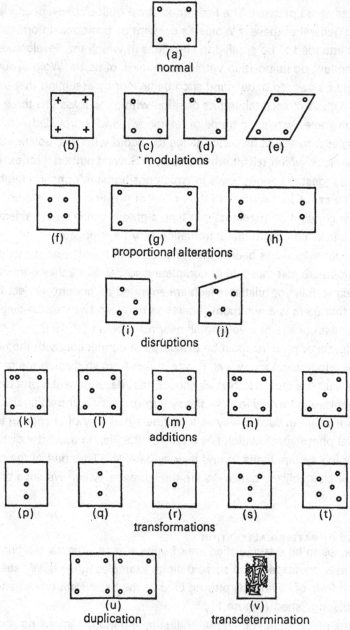

FIGURE 1
Categories of pattern alteration (details in text)

114

most animal patterns, for example, the four scutellar hairs in Drosophila are arranged *on the scutellum*. Wolpert's positional information is always information about position within, and related to, the pattern area.

The simplest alteration in a pattern is to change the elements, for example, turning some or all of the diamonds into spades (Figure 1b). Since we are dealing with patterns in developing organisms, such changes may occur by altering the course of development. For instance, each of the four scutellar bristles normally arises from a bristle-forming trichogen cell and a socket-forming tormogen cell. A gene such as *shaven-naked* causes the trichogens to develop as tormogens, so that the character of the four elements is changed. In *achaete* the alteration is more striking, in that some of the bristles on the mesothorax may completely fail to develop (Figures 1c and 1d). If such changes of development differentially affect certain elements in the pattern in a regular manner (for example, the anterior ones, or the posterior ones), then this might indicate that there is some previously unobserved factor in the pattern to account for this regularity; this we will call the *pattern field*, and will discuss it further below. But there is also the possibility that the pattern had remained essentially unaltered and the gene merely interrupts one of the processes of the differentiation of the elements. Stern [13] has in fact developed a method of preparing mosaic individuals of Drosophila which provide a complete demonstration that *achaete* does not alter the pattern of the bristle-forming sites on the scutellum, but merely interrupts the bristle development at one or more of them.

Another mild type of alteration—so mild indeed as scarcely to qualify as a change in pattern—is a coordinate transformation of the whole complex of pattern region and pattern arrangement (Figure 1e). (Mere changes of scale do not concern us here.)

Next one might consider proportional alterations of the arrangement of pattern elements, of the pattern region, or both. (a) In proportional alteration of pattern arrangement, the pattern region remains unaltered (or changed only in size), while the arrangement within it suffers a scaling up or down of coordinates (Figure 1f). (b) In proportionate alteration of pattern region, the shape of the region is changed, and the arrangement of elements changed 'to correspond' (Figure 1g)—the notion of 'correspondence' needs clarification, but we will not delay for that now. (c) In proportional alteration of both arrangement and region, the shape of the region is changed, and the arrangement is altered in some way which does not appear to be a *direct* effect of the regional change; for example, in Figure 1h the arrangement exaggerates the elongation of the region. Using

Form and information

Wolpert's terminology one might say that in (a) the framework for positional information remains the same, but the information is decoded differently ; in (b) the framework is changed, but the information decoded in the original way, while in (c) both framework and decoding are altered. In the Drosophila wing it seems to be easier to find examples of (a), for example, *shifted*, or (c), for example, *dachs*, than of (b).

It must be remembered that the pattern region at one stage of development, for example, the wing blade in Drosophila, was itself an element in a larger pattern at an earlier stage, for example, in the wing–thorax epithelium of the imaginal disc. A change of shape of pattern region (b or c) has therefore itself been derived from an earlier proportional alteration of pattern arrangement (a).

Disruptive alterations of patterns are characterized by replacing orderly systems of proportions by disorderly systems. The lack of order can be recognized, or perhaps one should say, can be defined, by the patterns being highly variable from individual to individual—if they are repeated identically in many individuals, they must incorporate some order, difficult though it may be to discern. Disruptions can affect only the arrangement of elements (Figure 1i) or both arrangement and region (Figure 1j). In the wing, some alleles of *cubitus interruptus* may belong to the former type, while *dachsous, comb-gap* and other genes certainly are examples of the latter.

We then come to the category of additions. The simplest types to understand are multiplication of existing elements (Figure 1k) and the addition of new elements in random locations (Figures 1l, 1m, and 1n). When we consider additions beyond this, we come into a borderland between this category and the next, most crucial one, which involves genuine topological changes of pattern. The important distinctions can be made only if these categories are compared with each other, so the second of them must be defined now.

Transformations of pattern are made when the arrangement of elements within the pattern region are topologically changed. Examples would be the transformation of a four of diamonds into a three, a two, or a one, or into a face-centred triangle, as in Figures 1p, 1q, 1r, and 1s. Transformations of this kind must involve a complete reorganization of the system of forces which bring about the arrangement of the element. They form, therefore, the most radical and interesting type of alterations of pattern. One cannot, however, always recognize *a priori* that a certain change is a transformation. For instance, is a change of a four into a five necessarily a transformation ? Might it not indicate merely that the original pattern of a four was based on some level of a precursor, varying continuously

116

over the region—rising above a threshold at the four corners, but only slightly below it in the centre, where the new element comes in ? In that case the change would not be a transformation, but would be an example of an addition revealing a previously unexpressed characteristic of the field of forces generating the pattern. We shall use the expression 'pattern field' for this underlying set of forces. The concept is similar to Stern's idea of a 'prepattern' [13], in that both notions postulate the existence of a distribution of some invisible precursor or substrate material ; but Stern's prepatterns seem always to be identical to the normal visible patterns, whereas the pattern fields are called in to account for the arrangement of new elements additional to those of the normal pattern.

Wigglesworth's hypothesis [see 10] that, in bugs, existing hairs or bristles are surrounded by zones of inhibition, so that new elements are placed as far as possible from any existing ones, amounts to the suggestion of one possible mechanism by which a pattern field might operate. When we discuss other pattern fields, for example, in Drosophila scutellars and wing-veins, we shall see that this mechanism does not operate in all circumstances.

In the case of the scutellars we find that when new elements are added, the first of these often appear near the existing bristles (as in Figure 1k). This suggests a rise of the pattern field still further above the threshold. When even more elements are added, they may appear in the middle of the sides, as though that region had now been brought up to threshold (as in Figure 1m) and sometimes we get Figure 1n.

Within the category of transformations, there is an important distinction to be made between 'organized transformations', which convert the pattern into another regular and orderly pattern (Figures 1p, 1q, 1r, and 1s) and 'disorganized transformations' in which the result is irregular and disorderly (Figure 1t). Since there is no obvious reason why any particular arrangement should be held, on a priori grounds, to be orderly, the recognition of order must be based on observation of repeatability and invariability between numerous examples ; one can, however, guess intuitively that clearly articulated geometries are more likely to be truly orderly than less easily comprehended ones, and experience tends to bear this out. (It is not always easy to distinguish disruptive alterations from disorganized transformations, and probably it is not very meaningful to try to do so in the case of some of the more messy and disorderly patterns.)

Another category of alteration is a complete or partial duplication of a pattern region (Figure 1u). As is well known, such duplications often exhibit phenomena of mirror-imaging. The old-established rules about this, formulated by William

117

Form and information

Bateson at the beginning of the century, have recently been discussed again in a very stimulating way by his son Gregory Bateson [14].

As a final category of alteration, one must consider changes which go beyond transformation of pattern, namely the conversion of the pattern region into a region of a totally other character, such as turning the four into a court card (Figure 1v), for example, the conversion of a wing into a scutellum, or an arista into a leg. These are best referred to by their classical name of homeotic changes; presumably the processes involved are similar to, or the same as, transdeterminations.

The studies on pattern formation in Drosophila have produced examples of nearly all these types of elementary alterations. They may occur at various stages during the developmental processes by which a final adult pattern comes into being, and when a certain alteration takes place early, there are likely to be consequential changes at later stages. Such consequential changes provide valuable evidence about the mechanisms by which the later-appearing elements in a pattern become located.

BRISTLE PATTERNS
The scutellar bristles. The scutellum of Drosophila is a trapezoidal plate of chitin forming the posterior section of the mesothorax. It normally bears four bristles or macrochaetae, one at each corner. The pattern is quite invariable in normal 'wild-type' or unselected stocks, even in extreme culture conditions, although occasional variants do occur. The pattern can therefore be said to be highly chreodic, or very well canalized. The scutellum actually develops by the fusion of portions of the two mesothoracic discs, and is therefore essentially bipartite in origin. There has been some controversy [15, 16] whether the canalization is better thought of as applying to each of the two sides (or even to the probability of occurrence of a bristle at an individual site), instead of in the more usual way as applying to the pattern as a whole. But this consideration hardly affects the later discussion, although it would call for some alterations in expression.

The scutellar bristle pattern has been extensively studied, but almost nothing is known of the developmental processes leading to the initiation of bristles in Drosophila. It is even uncertain whether they arrive at their final positions by migration from elsewhere [cf. 17], or, as seems more probable, arise by modification of certain cells in the hypodermis. The development of the *scute* or *achaete* phenotypes, which form the basis of Stern's notion of a 'prepattern' (*see* p. 124) do not seem to have been inspected since the early report of Lees and Waddington

[18], who made the not-very-illuminating observation that, at the very earliest stage at which bristle precursor cells can be identified in the hypodermis, *scute* individuals show certain sites at which the expected bristle precursors are absent.

If we cannot, for these reasons, form any definite idea of what the 'prepattern' is a distribution *of*, studies which have been concerned with change in the normal pattern of four bristles have led to information about the geometrical distribution, and the action of the factors which lead to bristle formation. We shall refer to the distribution of these factors as 'the pattern field'. The changes in pattern have been produced by selecting genetically different strains. The first worker to produce Drosophila with an altered scutellar pattern was Payne [19] ; the more refined recent work has been carried out mainly by Rendel and Fraser and their pupils [*summarized* 15, *also* 20-2].

The main concern of these authors was to establish and explore the fact that the scutellar pattern is chreodic or canalized. They compared, in quantitative terms, the amount of genetic variation contained in genotypes which yield a four bristle pattern with that, considerably smaller, in genotypes which yield the six, five, three, or two bristle classes. This is an important contribution to our general understanding of patterns, in line with the results of the studies of excess or deficiency of *cross-vein* mentioned on p. 129, but the details of this work would probably conventionally be classified as genetics rather than developmental biology. The main point of discussion for geneticists is to try to determine the relative importance, in altered phenotypes, of (a) changes in the quantity of some postulated 'bristle-inducing factor', as compared with (b) changes in the tolerance of the reacting cells to varying levels of the inducer. Rendel originally argued [23] that the genotypic response to selection was almost wholly by alterations of the former kind (changes in what he calls 'Make'), but it now appears probable that there may also be alterations of the second kind (changes in the degree of canalization). We shall not pursue these questions further here.

As a side issue to their main line of argument, the Rendel–Fraser studies have yielded considerable information about various aspects of the system which determines the scutellar pattern. This emerges when we examine the spatial distributions of bristles in flies showing numbers less or more than the standard four. Some drawings of such patterns are provided by Fraser [24], and fuller accounts are given by Sheldon [20] and Latter [22], who also provides more data on the frequencies of occurrence of the various types. There are some minor differences, arising presumably from the different stocks used, but the general

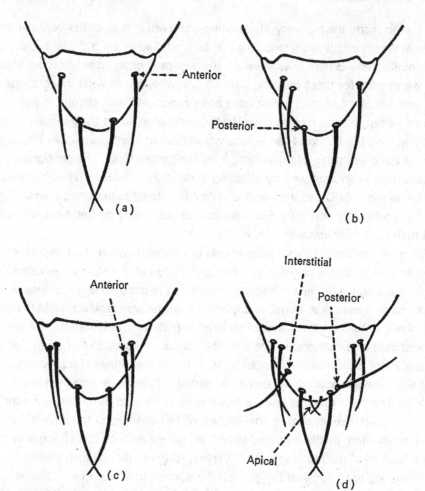

FIGURE 2
Patterns of scutellar bristles
The normal pattern is at the top left. The other figures show patterns in individuals from stocks selected for high numbers of bristles [from 22]

outlines of the shape of the pattern emerge fairly clearly. They can be summarized as follows:

(a) Bristles are formed almost exclusively on the lateral and posterior sides of the scutellum, and are very rare indeed on either the anterior or the middle of the scutellar area (Figure 2).

(b) Excess bristles are produced more easily, and bristles are lost with more difficulty in the anterior positions than in the posterior ones. Sheldon gives the following figures for variations in an unselected stock (raised under various

120

conditions) : in flies with 3 bristles, the missing one was anterior in 13, posterior in 16, and there were 25 with a reduced posterior bristle, none with a reduced anterior ; in flies with 5 bristles, the extra one was anterior in 108 cases, posterior in 1, and there were 54 with an 'interstitial' extra midway between anterior and posterior (Figure 3).

Latter [22] gives further data on the distribution of extra bristles in several different selected lines, and points out that although there are always more extras in anterior than in posterior positions, the exact ratio between the two varies between lines, and is therefore under genetical control.

(c) The favoured sites for bristle formation are regions rather than single-cell positions. Within such a region, there is no mutual inhibition such as that postulated by Wigglesworth to explain evenly spaced distributions. Bristles can in fact occur in closely packed groups of two, three, or four.

(d) The patterns suggest that there may be an inhibition exerted by one region on another. Thus the bristles which occur outside the standard four corner regions do so midway along the sides.

(e) The fact that one can find considerable variation in the patterns with five, or with six or more bristles shows that, in these, stochastic factors must play a considerable role in determining where a bristle is actually produced. If the processes were wholly determinate, each even bristle number would occur only in one pattern, and each uneven number only in two, depending on the greater development of one side or the other. This is far from being the case.

(f) When selection has raised the average numbers of bristles above about eight, a new factor enters the picture. Latter has shown [22] that, under these circumstances, there is a negative correlation between the numbers of anterior and posterior bristles. It is argued that this suggests that the groups of bristles are competing for some substrate substance whose concentration becomes limiting at these high-bristle numbers ; but since the number of interstitials remains positively correlated with the number of posteriors, the situation looks more complicated than this simple scheme.

Stern has discussed [13] bristle patterns, comparable to that on the scutellum, in terms of a 'prepattern'. He showed that there is some condition underlying the visible arrangement of bristles on the adult thorax of Drosophila, since even when one of the adult bristles does not appear because its site is occupied by tissue which inhibits its development (for example, homozygous *achaete* tissue), the location at which it would be expected still retains the property of forming a bristle if suitable cells are available. The *scute* gene, which is involved in many

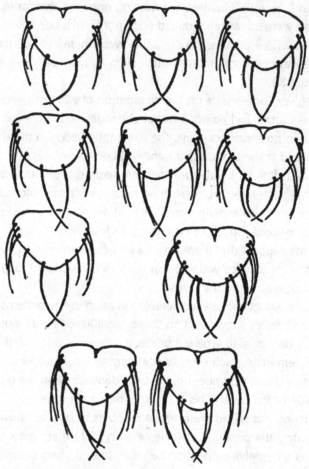

FIGURE 3
Patterns of increased scuttelar bristles
Further examples from stocks described by Sheldon [20]

of the stocks used by the Australian workers, presumably acts in this way, and although its effects on the scutellars are too random to be investigated by mosaics, it rather regularly removes an ocellar bristle, and this phenotype exhibits the same phenomena as those described for the thoracic bristles with *achaete* [25]. In such cases, there is no alteration of the pattern of potential bristle sites, but only an effect on whether the potentials are realized or not ; and thus they can give us no information about the factors determining the pattern except the bare fact, which is obvious on general grounds, that these factors involve a temporal sequence of processes. The Australian school have definitely succeeded in changing the pattern of the scutellars, and this presents us with a much more definite situation for which we are required to suggest an effective model. One might, for instance, propose that bristle formation has a certain chance of occurring when some precursor rises above a certain threshold ; that this precursor is accumulated around the position of a developing bristle (accounting for multiple bristles in one region), but is rapidly destroyed in regions away from the lateral edges of the scutellum (which accounts for the absence of bristles except along these edges) ; that the effective range of accumulation around a bristle area is less than half the distance between anterior and posterior sites (accounting for the appearance of interstitial bristles) ; and that the threshold of reaction to the substance increases from anterior to posterior (so that bristles are added more easily in the anterior, lost more easily in the posterior).

This would lead to a situation in the level of the postulated precursor and would have a form somewhat like the surface shown in Figure 4. The exact form of the surface must be under genetical control, since the relative frequency of extra anterior or posterior bristles is different in different selected lines.

Such a model is somewhat more complicated (because the phenomena to be explained are more complex and interesting) than the system of mutual

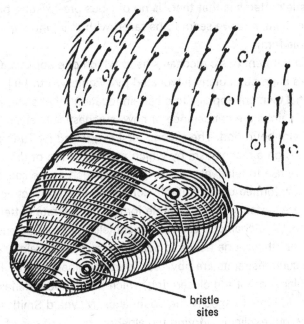

bristle
sites

FIGURE 4
Shape of the pattern field for the scutellar bristles

123

Form and information

inhibition, or competition for substrate, invoked to account for the even spacing of hairs ; but it is no less plausible *a priori*, and shares exactly the same lack of positive experimental support. One could doubtless invent other theoretical models which would account for the available facts equally well, but they would necessarily be of a similar order of complexity. The problem of accounting for a pattern even as simple as that of the scutellar bristles is not solved by invoking a prepattern with the same geometrical form, but demands that we find a way of explaining the production and maintenance of a distribution of morphogenetic substances of about this order of complexity.

▶ *Discussion of bristle patterns.* The easy way to explain the appearance of some biological pattern is to argue that it is merely an epigenetic elaboration of some pre-existing simpler pattern which was formed at an earlier stage. This type of hypothesis can plausibly be invoked for a great many, perhaps the majority, of the morphogenetic events in embryos, which one can refer back to the primary asymmetries of the ovum without much fear of conclusive refutation. But this is, basically, to shirk the issue ; which is, how does a system of chemical syntheses (the 'DNA–RNA–Protein' system) give rise to a pattern of specific elements arranged in a three-dimensional space ? One of the most challenging features of Drosophila bristle patterns is that there is no obvious pre-existing pattern from which to derive them, and we seem to be faced with this fundamental problem in an inescapable form.

There are two different basic mechanisms to which one appeals. One, which in recent times has been particularly advocated by Wigglesworth [9], is to postulate that an initial element appears in a certain location by chance, and that its presence there inhibits the appearance of new, stochastically arising elements in its immediate neighbourhood. The second, first suggested by Turing [26] (*see also* [27]), relies on a mechanism by which a combination of diffusion and reaction can give rise to relatively regular patterns, such as we see, for instance, in the Berard cells formed under suitable conditions in a convecting fluid layer.

Neither of these systems, in their simple form, can generate patterns of the orderly and repeatable character seen in the scutellar or thoracic bristles of Drosophila. To get these, one needs to add some further restraints. With Turing systems, some such restraints are obvious. Turing originally solved his equations for an infinite plane, or for the closed ring which had no boundaries. In any finite region, one would have to specify the boundaries. Maynard Smith and Sondhi claim [28], without explicitly quoting the algebra, that solutions of the Turing equations in a rectangular area give rise to orderly rows of spots, which they compare with the orderly rows of microchaetae on most regions of the Drosophila body. This gives a simple beginning, from which it would not be too difficult to proceed, by further elaborations, to account for such patterns as that of the scutellar pattern field.

It is less clear how one could approach this from the Wigglesworth theory, which, in its basic form, contains no reference, explicit or implicit, to anything other than random initiation of elements independent of the boundaries of the area. Moreover, the 'Wigglesworth mechanism' is based on a process (exclusion by a bristle of the formation of another in its immediate neighbourhood) which can certainly not be generalized; we have seen with the scutellars, and it holds with the head bristles also, that, although there may be an exclusion principle between bristle-forming regions, this does not occur within regions, where bristles may arise closely side by side. It seems likely therefore that, although the Wigglesworth exclusion process may be responsible for bristle distributions which show no more pattern than stochastic even spacing, the explanation of the more regular and precisely-located arrangements is to be sought in suitably constrained developments of Turing-type systems.

THE WINGS

The pattern elements in the adult wing are : the margin, which defines the pattern region, and bears a special series of hairs and other organules, the five longitudinal veins, and the two cross-veins. This form develops from part of the thickened epithelium of the mesothoracic imaginal bud. The wing portion of this, shortly before pupation, becomes folded into a two-layered sac. After pupation, this sac elongates both by cell division and by a thinning of the epithelium, into a tongue-shaped blade in which a first prepupal set of veins can be seen. There follows an extreme inflation of the sac by pressure of the internal body fluid : the epithelia are forced far apart and become very thin, the prepupal veins disappearing from view. Following true pupation (about 16 hours after puparium formation) the wing sac contracts again. The process involves the production of considerable tension in the epithelium, so that cuts or holes produced at this period by surgical operations become considerably enlarged [29]. By 48 hours the wing has assumed almost its final shape and venation. It is, however, still very small in area, and a final stage occurs in which the individual cells expand greatly.

The process of pattern formation (all facts cited in this section are from reference [11] unless one other reference is given) begins with the delineation of

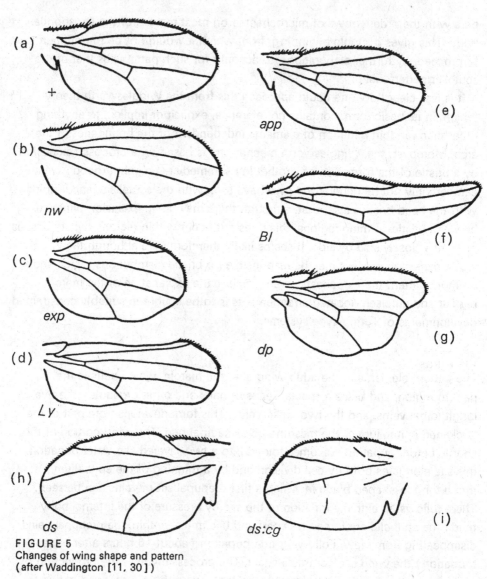

FIGURE 5
Changes of wing shape and pattern
(after Waddington [11, 30])

the wing region within the epithelium of the imaginal bud and its determination
as a region distinct from the mesothorax. This is probably a gradual process,
occurring mainly during the third larval instar. It is during this period that homeotic
transdeterminations and duplications may occur. Apart from such radical changes
we do not as yet have any good evidence of orderly transformations of the wing
into a topologically different pattern, comparable to the orderly transformations of

126

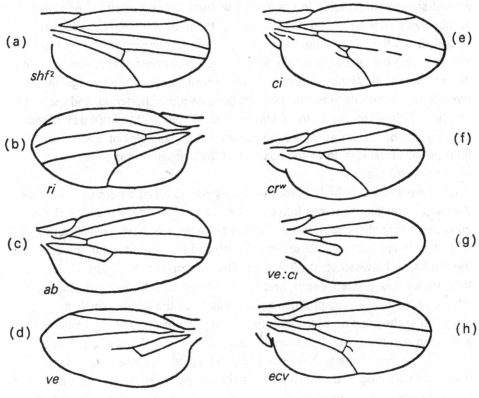

(a) shf²

(b) ri

(c) ab

(d) ve

(e) ci

(f) crʷ

(g) ve:ci

(h) ecv

FIGURE 6
Changes of the pattern of veins
(after Waddington [11])

the tarsus. However, there are several types of proportional alterations, both of the whole pattern region and of its elements. One rather surprising type of alteration involves shifting the relation between the borders of the region and the line along which the marginal hairs and so on will develop. The border of the wing region is originally defined by the line along which the epithelium of the imaginal disc becomes folded to produce the two-layer wing sac. Normally this coincides with the line of the marginal organules, but there are many genes located at several different loci, which shift one line in relation to the other, so that the wing develops with an abnormal shape, and with some regions of the margin missing (for example, Figure 5d). These alterations of the margin do not seem to have any primary effect on the positions of the longitudinal veins : they, like the margin itself, must have received their positional information from a

normal-shaped wing region in the disc epithelium before something occurred
(probably cell death) to shift the position of the fold which defines the definitive
boundaries of the wing blade. There are, however, secondary effects on the veins,
arising during the later contraction period, since the abnormal shape of the wing
disturbs the usual balance of tensions : thus unusually elongated wings become
more elongated at this time, shorter ones become more shortened, and the
absence of pieces of the relatively stiff anterior margin lead to local distortions.
These alterations could, therefore, be considered as examples of 'loss modulations', of a special kind in which it is part of the marginal areas of the pattern
region that are missing.

There are some orderly proportional alterations produced before or at the time
the wing epithelium becomes folded. In *shifted* (Figure 6a), the shape of the
wing blade is unchanged, but the longitudinal veins are shoved nearer together,
so that there are markedly fewer cells (which can be reorganized in the adult by
the cell hairs) between veins *L*3 and *L*4. This corresponds to Figure 1f. The
three genes, *four-jointed*, *dachs*, and *approximated*, which have very similar
effects on the tarsus, also produce similar effects on the wing, namely an
alteration of the position of the margin, to give a blunter shape, accompanied by
a splaying out of the veins, particularly toward their tips (Figure 5e). This
is an effect either of the type shown in Figures 1g or 1h, more probably the
latter. The widening of the angles between the longitudinal veins has a secondary
consequence on the formation of the posterior cross-vein, which is shifted
inwards, so that it lies nearer to the anterior cross-vein ; this effect was in fact
the origin of the name 'approximated'.

Disproportionate or disruptive alterations of the margin-vein system also occur.
Genes such as *dachsous*, *rotund*, and *comb-gap* (Figure 5h) give rise to wings
(with complete margins) of abnormal and variable shape, usually much broader
than normal, and with veins which are spread further apart (as usual in broader
wings), but in a disorderly and variable way [30]. These genes also have
variable and disorderly effects on tarsal segmentation (*see* p. 133). Whereas
genes of the first group show little interaction with one another in double
homozygotes—*d d app app ; fj fj app app*, or any of the other combinations
looking much the same as the single mutants—genes of the second group
show strong interactions in double homozygotes with any of the other genes.
Thus in *ds ds d d* ; or *ds ds cg cg* (Figure 5i) there is considerable overgrowth
of the thorax as well as of the wings and legs and venation, tarsal segmentation
and thoracic bristle patterns are all profoundly disrupted.

During the next developmental stage, the prepupal, the wing blade consists of two initially rather thick epithelia, which are in contact by their inner surfaces, except along certain lines which form the prepupal veins. The blade gradually expands in area, both by thinning and stretching of the epithelium and by cell division. A number of proportionate alterations can be made to the pattern at this time. The genes *broad* and *expanded* (Figure 5c) cause the wing to become broader and shorter, while *narrow* (Figure 5b) and *lanceolate* have the opposite effect. The alterations seem to be caused by control of the directions of mitotic spindles in the dividing cells.

Much more drastic coordinate transformations are characteristic of the next stages of development. At the end of the prepupal period the wing becomes blown up, by pressures of haemolymph, into a large hollow sac. The two epithelia maintain only tenuous contact with one another and the prepupal veins disappear. After a time, a contraction begins which brings the two epithelia back together so that they become connected by cellular processes everywhere except along certain lines which become the veins. The connections between the epithelia, which define the veins, are found first near the distal tip and extend along the anterior and posterior regions towards the root of the wing, the last region to be 'deflated' being that of the posterior cross-vein.

The contraction of the wing operates fairly slowly over a period of some hours. It involves a considerable tension in the epithelium, which is resisted by the internal fluid, and the developing veins and margin. As mentioned earlier, if parts of the margin are missing, the contracting epithelium can squeeze the whole wing into a different shape. Even in the normal wing, with complete margin, the contraction brings about some change of the whole outline, converting the lax bag-shaped prepupal wing into the tenser form of the adult wing. The balance of tensions and resistances must be rather precarious since there are a number of genes, of which *dumpy* and its many alleles are the best known, which cause the wing to collapse into an over-contracted form. Usually the over-contraction is along the proximo-distal axis (Figure 5g), but sometimes it can be in the direction perpendicular to this ; it often is so in the mutant *Blade* (Figure 5f), but the difference between the two forms must turn on very slight features, since one can find flies with a dumpy-type wing on one side and a Blade-type on the other. Dumpy and Blade (Figures 5f and 5g) distortions can also easily be produced as phenocopies by heat shock, and their frequency can be increased by selection [31]. These changes in wing pattern appear some of the most drastic on casual visual inspection, but it is clear that they are nothing more profound than elastic distortions of the basic pattern, and therefore of little interest except as examples warning us of the desirability of studying the development of a form before offering an explanation of it.

The changes in vein pattern occurring at this time are considerably more interesting. First, losses. The characteristic form of these is that the pupal veins first appear as normal, but then parts of them become obliterated by the coming together of the two epithelia at places where they should remain apart. This happens very clearly to the tips of the longitudinal veins in *veinlet* (Figure 6d) and *abrupt* (Figure 6c) and to a section of the fourth longitudinal vein in *tilt*. The end result is a normal pattern from which something has been deleted. This is a phenomenon of the same kind as that in the *achaete* studied by Stern (cf. p. 121), but whereas Stern had to employ special methods (involving somatic mosaics) to demonstrate that *achaete* does not alter the pattern but only the processes arranged in the pattern, in *veinlet* and *tilt* this follows from a simple inspection of their development. Stern's experiments also showed that in *achaete* and several other genes tested this deflection of development is carried out quite locally, by small areas of tissue. It is not known whether this is true in *veinlet* and *tilt*. In contrast to this situation in *veinlet*, the part of L2 absent in *radius incompletus* does not ever put in a visible appearance.

There are other genes which cause the appearance of extra vein material. This is usually not organized into coherent linear veins, but occurs as scattered fragments in the wing surface, often attached to some part of the normal venation (for example, the extra cross-vein in Figure 6h). The extra vein material appears whenever the contraction fails to bring the upper and lower epithelia into effective contact. With many genes, this function seems to be due to deficiencies in the contractibility of the epithelium itself, but in some cases (for example, *blistered*) the epithelia are held apart by entrapped globules of body fluid.

The most interesting aspect of the gene actions which cause loss or addition of veins is the relatively precise location of their effects. *Veinlet* removes the tips of the longitudinal veins, *tilt* affects specifically a central stretch of L4, and *cubitis interruptus* (whose effects begin to be visible on the prepupal vein) has a series of alleles each with rather localized effects in L3 and L4. *Net* and *plexus*, which add vein material, tend to do so in rather specific regions, but there are some other 'vein-adding genes', for example, *blistered*, whose effects are less localized and may occur anywhere in the wing. All these cases can be considered as changes in the wing in which positional information is decoded. The fact that there are so many different specific regional effects is evidence that there are many factors affecting the 'code-book'. In general we have little indication of the exact nature of these factors, although what we know of the developmental processes involved—that is, the formation of veins during the coming together of the two epithelia, during the contraction, suggests that physical factors of the contractibility of cell surfaces, and tendencies to form intercellular processes and adhesions, play an important role in determining the 'code'. Lees [29] found that the amount of extra wing material in *plexus* is reduced following injuries to the early pupal wing bud, presumably because of a release of internal tension.

The pattern element whose location we know most about is the posterior cross-vein. This is the last element to appear, since it arises as the persistent remnant of the cavity of the inflated wing. Its position is very much affected by many of the previous events of wing development. If the veins $L4$ and $L5$ are moved for any reason, the posterior cross-vein shifts into a position which tends to keep its length normal ; that is, if the longitudinal veins are splayed apart, as in *approximated, expanded*, and so on, the cross-vein moves inwards ; if they are pressed together, as in *shifted*, it moves outwards (Figures 5c, 5e and 6a). The effect is produced even by alterations which occur to the longitudinal veins at later stages, for instance by *dumpy* during the contraction phase (Figure 5g). When parts of the longitudinal veins are missing, the cross-vein shifts so as to retain contact with the parts that are left, for example, in Figures 6c and 6d. There is often, but not always, a smoothing-out of the angles between the cross-vein and any other small detached piece of vein which does not form part of a long continuous stretch. This can be seen in some examples of *cubitus interruptus*. With one allele of that locus, ci^w, one often finds a smooth vein formed by a fusion of the base of $L3$ through the cross-vein with the tip of $L5$ (Figure 6f). This pattern is repeated rather often, as though it had some internal stability or canalization. It is one of the venation changes which might possibly be considered as an organized transformation.

The cross-vein can also appear in a defective form, with a gap either in the middle or at one of its ends. It may indeed fail to appear (or to persist) at all (Figure 7a). Such changes are easily produced by heat shocks applied during the pupal period, shortly before the time when the vein becomes definitely formed [32-4]. Additions to the vein in the form of lateral branches may also appear (Figure 6h). Stocks have been built up [35] in which, under normal conditions, gaps or additions appear in various percentages of a population. The differences between the stocks were dependent on many genes. By using various crosses and backcrosses between these stocks, it was shown that

131

the range of genotypes which produces the normal phenotype, with a complete cross-vein, is much more extensive than that which produces a particular grade of deficiency or excess. The normal phenotype is, in fact, buffered or canalized, and can absorb, without noticeable deviation, a good deal of genetic variation, whereas the abnormal phenotypes lack this stability and vary readily whenever the genotype is altered. Rendel, Fraser, and their colleagues have since made many similar and even more thorough studies on the canalization of the numbers of scutellar bristles (cf. p. 119).

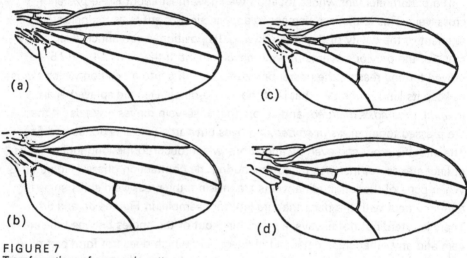

FIGURE 7
Transformations of cross-vein pattern
(after Bateman [36])

The use of temperature shocks made it possible for Bateman [36] to build up, by selection, a number of genotypes which determine vein alterations which, strictly speaking, must be classified as organized transformations, in that they regularly bring into being a pattern which is topologically different from the normal. When a temperature shock is applied to pupae of a normal laboratory stock of Drosophila, a number of modified phenotypes appear, including not only a loss of the particular cross-vein, but the appearance of extra cross-veins near the anterior one (Figure 7). By breeding from individuals showing one specific type of alteration, Bateman was able in a few generations to bring about the genetic assimilation of the effect, and to produce strains in which the phenotype appeared in all individuals under normal conditions, without any application of a temperature shock. These are genuinely new venation patterns, and it is perhaps surprising to find that they are produced by polygenic differences from the normal

132

genotype, while no single genes have been found with comparable effects. Their importance as potential evolutionary transformations of wing venation would be greater if they occurred at an earlier stage of development, so that they themselves produced secondary consequences, but in any case they are interesting models of ways in which topologically distinct transformation of patterns may occur.

THE LEGS

The main interest of the Drosophila leg, for the study of pattern formation, is that the various genetically produced alterations include some of the clearest examples we have of topological transformations, both into organized and disorganized forms. There can be distortions of the normal pattern, arising for instance during the contraction phase [37], but the most interesting pattern changes concern the segmentation of the tarsus [30, 38, 39]. Normally there are five tarsal segments. Several genes are known which reduce the number (none are known to me which increase it) (*see* Figure 8). One group of these, *four-jointed* (*fj*), *dachs* (*d*), and *approximated* (*app*), have very similar effects, causing the appearance of shortened legs with four regular tarsal segments : these genes also have very similar effects on the wings, producing well-organized proportional alterations affecting both the shape of the wing and its venation. Other genes (for example, *aristopedia* (*ss^a*), *dachsous* (*ds*) and so on) affect the tarsi in more irregular fashion, producing irregular tarsi, variable both in number and shape of the segments. There is sometimes some tendency for a particular region of the tarsus to be most strongly affected, for example, in *eyeless-dominant* the proximal part is the most altered, and weak expressions of the gene may show no more than an enlargement or disarrangement of the sex combs.

The two groups of genes behave very differently in compounds. Double homozygotes of *d*, *fj*, and *app* have about the same phenotype as the single homozygotes. Compounds between any of these and such disorganizing genes as *dachous*, *rotund*, *comb-gap*, *aristopedia*, or *eyeless-dominant* often show extremely exaggerated effects ; these include considerable overgrowth, not only of the legs, but of the head and thorax, and transdetermination-like changes of one organ into another.

The segmentation of the tarsus has, of course, profound secondary effects on more minor elements of the pattern, such as the position of the sex combs, bristles, and so on. These may well depend on specific development effects of the intersegmental membranes [10] but nothing definite is known about this in Drosophila.

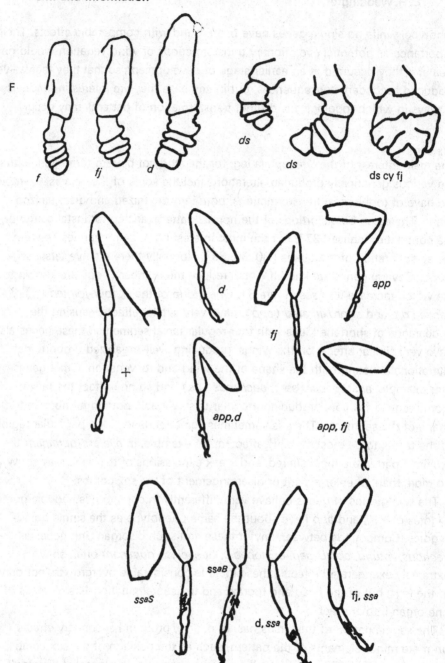

FIGURE 8
Patterns of tarsal segmentation
Above are drawings of the imaginal buds shortly after emersion, below outlines of adult legs
(after Waddington [1])

C.H.Waddington

Stern and Tokunaga [40] showed that the effect of *eyeless-dominant* in dis-
arranging the sex comb is not autonomous in mosaics, as all the other pattern
effects they studied had been. They concluded that this gene genuinely does
alter the 'prepattern' and not merely interfere with its expression, as *achaete* and
other genes had done. It would indeed be surprising if genes which produce a
topological transformation of a series of five into a series of four should prove to
act autonomously in a small region, since such a transformation would seem to
call for a complete reorganization of the way of decoding positional information ;
it is difficult to see how this could be effective locally.

Another case of a mosaic phenotype involving leg-like organs is perhaps
worthy of a slight digression. Roberts [41] found that in antennae mosaic for
wild-type and *aristopedia* tissues, each tissue developed autonomously into
either antennal or leg organs. He expressed considerable surprise at this, although
it has long been known that similar mosaic phenotypes are produced by some
alleles of *aristopedia* (for example, ss^{aB}) which convert part of the antenna into
leg structures while leaving the remainder antennal [11, 42-4]. Roberts interprets
his results by stating that 'the mutant and wild-type alleles of a homeotic mutant,
aristopedia, do not control "antennal" prepatterns, but control the competence
of antennal cells to respond to an "appendage" prepattern'. This usage surely
removed from the word 'prepattern' the last traces of its never very impressive
significance. If there is no difference between the prepattern of an antenna and
a leg, but they are both, in Roberts' words, 'an appendage prepattern which is
common to all imaginal discs', then obviously such a prepattern cannot give us
any help in understanding the origin of the pattern differences we see between
antennae and legs, let alone between the first, second, and third legs. It becomes
quite irrelevant to the basic questions of pattern formation, such as the arrangement of particular types of bristles or other elements at definite spatial positions.
The situation is better expressed in the classical embryological terminology, in
which one would say that *aristopedia* affects the determination of cells to become
either antennal tissue or leg tissue ; in fact it brings about a transdetermination.
One can then reserve the word pattern for the phenomenon to which it truly
belongs, namely the spatial arrangement of entities within an organ of a certain
type, for example, bristles on legs or aristal branches on antennae.

The temptation to use highly abstract notions, such as 'prepattern' may become
somewhat less if we consider the actual process of segmentation. The leg is
originally an area of epithelium, roughly circular in outline, on the inner surface of
an imaginal bud. This region increases in area, and as it does so folds out into a

135

cylindrical process. The tarsal region is at the distal tip of the process (Figure 8, *above*). It rapidly becomes constricted by a number of grooves which are arranged transverse to the long axis of the cylinder, and these become inter- segmental junctions : similar grooves are just what one would expect in a region of material which is growing in size and attempting to expand against some con- stricting boundary. There seems to be no reason to look for any factors other than the growth, extensibility and resistance to deformation of the epithelium to account for the initiation of the pattern of grooves. Their later development certainly involves other factors, including the assumption of special cell shapes in the corners. The grooves can, in fact, be envisaged as the results of simple hydrodynamical-elastic processes. It is difficult to think of an exact model in ordinary experience, where we much more often encounter things which are contracting rather than expanding : but one might say that the process of tarsal segmentation is like the *inverse* of the formation of concentric ridges on the surface of a congealing and cooling circular drop of candle-wax.

If the material in such a system is homogeneous, the grooves will form complete circles, and will not, for instance, connect with one another to form a spiral or other continuous curve. The spacing and number of grooves will depend on the properties of the material, including its rate of growth ; but there must be an integral number of them. Genes like *four-jointed* (fj) which give organized transformations, must change the cell properties in such a way as to preserve the general nature of the system while altering a few of the parameters. Genes like *dachsous* (ds), *rotund*, and *comb-gap*, which produce variable numbers and variable spacings of grooves, must change cell properties in ways which de- stabilize the dynamics of the system. It is impossible at present to categorize in any more definite way the difference between these two classes of cellular alterations : conceivably one affects cell membranes, the other growth rates, or something of that sort. It is a noteworthy, but not easily interpretable fact, that the 'organized' group (fj, d, app) have effects which, to normal observation at least, are confined to the wings and tarsi, while many of the 'disorganized' group (ds, rt, ey^D, ss^a) affect either the whole body (first two) or at least several diverse organs of it (last two).

GENERAL DISCUSSION

The work on Drosophila patterns described above has two particular features. In the first place, we are brought in contact with many different types of patterns : sequences of elements in a one-dimensional array (tarsal joints), arrangements of joints (bristles and so on), or of lines (veins) on a plane ; or shapes of plane regions (wing shapes). Secondly, we have studied rather a large number of ways in which these patterns may be altered, more or less radically. Nearly all these alterations have been caused in the first place by changes in the genotype of the cells ; that is to say, by alterations usually to one species of molecule, the primary protein produced by the gene (a few pattern genes with very general effects—for example, *dachsous*—might possibly be 'control genes', but it seems very unlikely that many of them are). The evidence available so far from Drosophila studies provides little information about effects on patterns produced by direct experimental surgery, such as experiments described by Lawrence in *Hemiptera*. But several of the primary genetic effects are followed by secondary consequences, for example, scalloping of the wing margins by changes in the reaction to the wing contractive forces, and in such cases we obtain much the same information as we would have done if the primary effect had been produced by a scalpel instead of by a gene.

The main advantage of the genetic method of studying pattern development is that it allows us, by changing just one of the chemicals in the reacting system, to make much more subtle and more profound modifications than can be achieved in any other way. Its disadvantage is that we cannot, as yet, identify the chemical we have changed by anything more illuminating than the conventional name of the gene. Undoubtedly our ideas about pattern formation in animals in general must have a wider basis than Drosophila, but the variety of facts and phenomena known in that insect allows us to make at least a start at formulating a general picture and seeing whether from this we can extract some simple but essential problems.

It is clear enough that all the patterns we see on an adult insect have been produced by sequences of developmental processes. For a pattern as a whole (such as the wing) we know of many factors which intervene at several stages in these sequences. Even a single element may have its position determined by several processes acting at different stages, as the position of the posterior crossvein reacts to the shapes of the wing, the scalloping of its edges, its mode of contraction, the original splaying or squeezing together of the veins, their persistence or disappearance in its neighbourhood, and so on. One might attempt to extract a simplified paradigm out of such situations along the lines suggested by Stern [13], involving a 'prepattern' and a 'competence' to react to it. But, as we have seen, the facts force us to envisage these factors as considerably more complex than Stern imagined. The 'competence' is not just a matter of 'react' or

Form and information

'fail to react' : a part of a vein may appear (in response to the 'prepattern') and then disappear again (as in *veinlet, tilt*), and not only would we have to postulate a sequence of prepatterns, but they have to have a different spatial arrangement than the visible normal patterns, in order to provide places in which additional elements may appear (as in scutellar bristles, or extra cross-veins).

We have used the expression 'pattern field' in this connection. Can this perhaps provide a paradigm concept ? It seems doubtful. A pattern field is a disposition of developmental forces or processes, and its material substratum must be very different in different cases. The forces which determine the probabilities of appearance of extra cross-veins in the wing are concerned with the mechanisms of deflation of a contracting elastic bubble. The forces which control the clustering together of extra scutellar bristles must be quite different, and they are also clearly different again from the process which results in the spacing apart of additional hairs in Wigglesworth's examples. We do not yet know much about the chemical systems involved in any of these examples, but if we did, for example, identify the actual 'excluding substance' in the Hemipteran hairs, the argument above suggests that this would not provide the fundamental new insights which some people expect. We should have further information about one process involved in a certain pattern field ; this would, of course, be interesting, but there would be no reason to think it could be generalized to apply to any other pattern field, let alone all.

One should, in fact, realize, as embryologists working in more classical systems have done for many years [cf. 45-7], that any concept of a 'field' is essentially a descriptive convenience, not a causal explanation. It may be convenient to speak of, or even draw a map of, the scutellar bristle field, or the cross-vein field, but the operative forces have to be experimentally identified separately in each instance : only if the forces are always the same or of a very few kinds, as they are in gravitational or electromagnetic fields, or if the maps were always the same, would the field concept be a unifying paradigm ; and we know that none of these conditions is fulfilled.

Using Wolpert's terminology, a pattern field can be regarded as a set of positional information together with a code for interpreting it into developmental processes. The most important tasks in the study of patterns at the present time would seem to be to attach more definite meanings to these two notions. Until recently, there was little to be discussed in a meaningful way about the nature of the positional information, although as Wolpert has emphasized [2], there are many problems about how much of it is required. Only one hypothesis was

138

available, namely that position is specified by gradients. One might ask, what are they gradients *of* ? But many biologists had realized that the answer was likely to turn out to be quite trivial. Cells, each with a highly complicated metabolism, involving control mechanisms at several different levels, may communicate with one another by a gradient in some low-level metabolite (even an inorganic ion) whose nature will throw very little light on the essential processes taking place at or near the gene level. It might reveal no more about the essentials of the system than the identification of the medium of communication as paper would tell us about the nature of a bureaucracy or business organization. In the last few years an alternative to gradients has been proposed, namely the phase-shift model of Goodwin and Cohen [5]. This has made the whole question a very active one again, and one of the major immediate tasks is to determine whether one of these two models is the more basic, or whether some patterns are based on gradients, others on a phase-shift system. Again, the question of identifying exactly what is oscillating in a phase-shift system seems of less importance than discussing whether that kind of system is employed at all.

The major problem in this area concerns the production, by elementary altera-tions in the pattern-generating systems, such as those produced by single gene mutations, of topological changes of the final phenotypic patterns. It is clear that changes of this kind, both orderly and disorderly, can be produced in one-dimensional patterns ; there are clear-cut examples in the patterns of tarsal segmentation. It is the situation in two and three dimensions which presents the challenge. In theory, such changes might arise in patterns which are produced by a Turing-type mechanism. Maynard Smith and Sondhi have pointed out [28] that the topological arrangement of such patterns is dependent both on the position of the boundaries of the area concerned, and on the various parameters of the diffusion and reaction rates involved ; a change in any one of these might bring about a change from one topological pattern into another. Gmitro and Scriven have discussed and illustrated [27] a number of different elementary pattern formations both in the plane and on the surface of a sphere. However, we still lack any very convincing examples of such phenomena among the many pattern alterations which have been described in Drosophila. The formation of additional cross-veins is perhaps the best case, but, as was pointed out, this arises at such a late stage, and seems to depend on such a special mechanism of the physical contraction of a membrane-bound bubble, that it may be considered somewhat trivial. Alterations as far reaching as topological changes would be expected to occur less often than such minor changes as modulations,

Form and information

proportional alterations, and additions. But perhaps they are not in fact as rare as the existing data would suggest. The urge to search out, for study, systems which are as simple as possible, has probably led to an undue concentration on one-dimensional systems, which are simple enough ; indeed so simple that results obtained with them cannot necessarily be generalized to throw light on systems involving more dimensions.

References

1. C. H. Waddington, *New Patterns in Genetics and Development* (Columbia University Press 1962).
2. L. Wolpert, in (C. H. Waddington, ed.) *Towards a Theoretical Biology 1: Prolegomena* pp. 125-40 (Edinburgh University Press 1968).
3. L. Wolpert, in (C. H. Waddington, ed.) *Towards a Theoretical Biology 3: Drafts* pp. 198-228 (Edinburgh University Press 1970).
4. L. Wolpert, in (C. H. Waddington, ed.) *Towards a Theoretical Biology 4: Essays* pp. 83-94 (Edinburgh University Press 1972).
5. B. C. Goodwin and M. H. Cohen, *J. theoret. Biol. 25* (1969) 49-107.
6. C. H. Waddington, in (C. H. Waddington, ed.) *Towards a Theoretical Biology 2: Sketches* pp. 179-83 (Edinburgh University Press 1969).
7. F. H. C. Crick, *Nature 225* (1970) 420-2.
8. M. H. Cohen, *Symp. S.E.B. 25* (1971).
9. V. B. Wigglesworth, *J. exp. Biol. 17* (1940) 180-200.
10. P. A. Lawrence, *Adv. Insect. Physiol. 7* (1970) 197.
11. C. H. Waddington, *J. Genet. 41* (1940) 75-139.
12. C. H. Waddington, *The Strategy of the Genes* (Allen and Unwin : London 1957).
13. C. Stern, *Amer. Sci. 42* (1954) 213-47.
14. G. Bateson, *J. Genet.* (In press).
15. J. M. Rendel, *Canalisation and Gene Control* (Logos Press : London 1967).
16. A. Robertson, *Amer. Nat. 99* (1965) 19-24.
17. H. Ursprung, in (M. Locke, ed.) *Major Problems in Developmental Biology* pp. 177-216 (Academic Press : London 1967).
18. A. D. Lees and C. H. Waddington, *Proc. roy. Soc. B 131* (1942) 87-110
19. F. Payne, *Proc. natn. Acad. Sci. Wash. 4* (1918) 55-8.
20. B. L. Sheldon, *Aust. J. biol. Sci. 21* (1968) 721-40.
21. A. S. Fraser, in (C. H. Waddington, ed.) *Towards a Theoretical Biology 2: Sketches* pp. 184-203 (Edinburgh University Press 1970).
22. B. D. H. Latter, *Genet. Res. 15* (1970) 285-300.
23. J. M. Rendel, *Evolution 13* (1959) 425-39.
24. A. S. Fraser, *Genetics 48* (1963) 497-514.
25. C. Stern and D. L. Swanson. *J. Fac. Sci. Hokkaido, Series VI, Zoology 13* (1957) 303-7.
26. A. M. Turing, *Phil. Trans. roy. Soc. B 237* (1952) 37-72.
27. J. I. Gmitro and L. E. Scriven, in (C. H. Waddington, ed.) *Towards a Theoretical Biology 2: Sketches* pp. 184-203 (Edinburgh University Press 1969).
28. J. Maynard Smith and K. C. Sondhi, *J. Embryol. exp. Morph. 9* (1961) 661-72.
29. A. D. Lees *J. Genet. 42* (1941) 115-42.
30. C. H. Waddington, *J. Genet. 51* (1952) 243-50.

Comment by René Thom

31. N. G. Bateman, *J. Genet. 56* (1959) 341-52.
32. C. H. Waddington, *Evolution 7* (1953) 118-26.
33. R. D. Milkman, *Molecular Mechanisms of Temperature Adaptation* pp. 147-62. (Amer. Assoc. Adv. Sci. 1967).
34. R. D. Milkman and B. Hille, *Biol. Bull. 131* (1966) 331-45.
35. C. H. Waddington, *Zeits ind. Abst. Vererblhr. 87* (1955) 208-28.
36. N. G. Bateman, *J. Genet. 56* (1959) 443-54.
37. C. H. Waddington, *Proc. Zool. Soc. A. 111* (1941) 181-8.
38. C. H. Waddington, *Growth Suppl.* (1940) 37-44.
39. C. H. Waddington, *J. Genet. 45* (1943) 29-43.
40. C. Stern and C. Tokunaga, *Proc. natn. Acad. Sci. Wash. 57* (1967) 658-64.
41. P. Roberts, *Genetics 49* (1964) 594-8.
42. C. A. Villee, *J. exp. Zool. 93* (1943) 75-98.
43. M. Vogt, *Biol. Zent. Bl. 65* (1946) 238-54.
44. C. H. Waddington and R. M. Clayton, *J. Genet. 51* (1952) 123-31.
45. C. H. Waddington, *Principles of Embryology* (Allen and Unwin : London 1956).
46. C. H. Waddington, in (M. Locke, ed.) *Major Problems in Developmental Biology* (Academic Press : London 1967).
47. P. A. Weiss, *Principles of Development* (Holt : New York 1939).

Comment by René Thom

I read your article on pattern formation with great interest. As for the points you discuss, the most interesting part seemed to me your criticism of the 'field' concept. In pages 138 and 139 you write two paragraphs which seem to me contradictory in spirit if not in words.

In the first, you say that the field concept has just descriptive value, and no explanatory power. To get a true explanation, one has to know the nature of the 'operative forces'. 'Only if the forces were always the same or of a very few kinds, or if the maps were always the same, would the field concept be a unifying paradigm ; and we know that none of these conditions is fulfilled.'

I have very strong objections against this reductionist and anthropomorphic viewpoint. At the finest level of analysis, undoubtedly Wolpert considers, as one of the best explained morphogenetic situations, gastrulation in sea-urchin, where he can see the agents, the mesenchyme cells, with their pseudopodal activity. It never occurred to him that the problem is not solved, but just a bit displaced when you identify 'local agents' : why do mesenchyme cells have, precisely at this time, this strange activity ? Why is it that this activity ceases when the two tissues have been put into contact ? And so on. . . .

If water freezes below 0°C, do you know the 'operative forces' which bring about this morphological change ? Molecular interactions, you may say. But are not these forces already working above 0°C ? So the same forces may bring

141

about, according to some change of external parameters, very different morphological appearances. Moreover, even when the morphogenetic forces are 'intuitively' obvious, as when a compressed sheath of tissue (or cloth) folds itself, the process may be nevertheless highly unstable, and very difficult to formalize, mathematically or physically. For these two reasons, I believe it would be safer to forget about forces and to replace them by their—mathematically more or less well defined—potentials, and try to get at the purely formal aspect of the regulation in fields or chreods. This is the point you stress in the following paragraph, where I feel, of course, in full agreement with what you say.

As for the multiplicity of fields models, one has to be a bit careful. I believe that local accidents, in a pattern or in a field, belong only—in general—to a finite catalogue of chreods (otherwise, you could not describe them verbally) ; it is their association in global patterns which, of course, leads to infinitely many models. But this does not preclude any usefulness to the 'field' concept. Would you say, in linguistics, that the concept of *sentence* is useless, because there are, in a language, infinitely many sentences ?

Reply by C. H. Waddington

Your main criticism, about my discussion of 'fields', raises a very interesting and, I think, difficult point. When I said that 'any concept of a "field" is essentially a descriptive convenience, not a causal explanation', I was not expressing myself very well. There is, I suppose, in the last analysis, no distinction between a description and an explanation in terms of postulated 'operative forces'. But I went on to allude to the fact that in certain types of situation we are usually ready to accept a certain 'description' as adequate for the time being to count as an 'explanation' behind which we do not feel urgently called upon to enquire, while in other circumstances the best description we have of certain phenomena do not satisfy us in this way. For instance, we feel reasonably satisfied to work with the idea of an electromagnetic field because, although the field may have many different geometrical forms in different cases, the variable mapped is always 'electromagnetic potential'. (Of course, we have no idea what electromagnetic potential *is*, but we are usually satisfied to adopt the view that such a question would be meaningless, until someone like an Einstein shows us how electromagnetism can be connected with other basic constituents of the universe, such as matter, energy, and so on.) Again, we accept gravitational fields, although

142

Comment by Lewis Wolpert

gravity is obviously even more mysterious than electromagnetism, because the main examples of the field have basically similar geometries. But if one thinks of some examples of epigenetic fields, such as those of the pentadactyl limb, the dogs' skulls or fish-shapes which D'Arcy Thompson put into his coordinate nets, the Drosophila scutellar bristles, or the regenerating Hydra, not only are the geometries all different, but one cannot help asking oneself whether, perhaps, one is a diffusion field of some substance or substances, another a phase-shift field à la Cohen–Goodwin, another perhaps produced by interaction between growth-centres, and so on. Merely to say of such a phenomenon that it involves a field, does not 'satisfy' one, and remove the temptation to ask the next question.

This does not mean that the concept of fields is useless in biology. I think that at the end of your letter you raise a very useful analogy when you ask whether the concept of a sentence is useless in linguistics because there are infinitely many sentences. Of course, it remains a very useful concept ; but it does not amount to a 'causal explanation' of how the words of a particular sentence are arranged. That 'explanation' is provided by a grammar. The point I am making about fields is, that to say a developmental performance involves a field is as important as saying that a collection of words is a sentence ; but just as we need to understand the grammar which generated the sentence, so we need to enquire what generated the field.

Comment by Lewis Wolpert

I think this a very interesting paper because it does, as you point out, summarize an enormous amount of important information. It is perhaps only natural that I should disagree with some of your major conclusions and I think this is partly because we do not have the same concept of positional information, or the relationship between pattern and morphogenesis. My own feeling is to follow your original terminology and to talk about morphogenesis, or the development of form, as being principally concerned with the cellular forces that give rise to change in shape and to think about pattern formation, as you did, as the assignment of differences to a group of cells such that they develop a well-defined pattern. I think that one should be quite careful about the distinction between these two processes, since, for example, one group of cells surrounding another clearly forms a pattern, but this could result from morphogenetic processes such as the cells sorting out as in a typical Steinberg experiment, or could be a pre-

143

pattern phenomenon in which the cells are assigned different states without movement. I am thus not in full agreement with your analysis on page 115, and particularly with your suggestion that the most striking feature of pattern is a number of discrete elements arranged in some form or order. I would certainly speak of the pattern of a femur or a scapular, and it is exactly in this sense that I would think about it. However, I think that one should concentrate on the process rather than the final design, because I think if one does not do this, one can be very misled as to what is happening. For example, in gastrulation in the sea urchin, the specification of the mesenchyme cells is a pattern process, whereas the pseudopodal activity, which brings about a considerable change in form, is a force phenomenon. My concept of positional information, or, if you prefer, gradient, is that one does not have to think that pattern reflects some level of a precursor as you suggest on page 123, and again on page 138. On the contrary, my whole idea is that in order to generate a pattern, one does not require to generate a prepattern, as Stern thought. One only needs to have some measure of relative position, which can be provided by a gradient. It is no surprise to me that all Stern's prepatterns are the same, because I would argue that there is absolutely no relationship between the form of the gradient and the pattern expressed when it is interpreted by the cells. There is thus no reason to get involved with maintaining a complex distribution of morphogenetic substances, nor need one get involved with the sort of ideas that Maynard Smith, and Turing, and Scriven have put forward, which effectively provide pattern gradients. To my mind these are simply not necessary.

As regards the eyeless dominant mutant of Stern (page 135), because of the break in the intersegmental membrane, one does not have to invoke a true change in prepattern. Moreover, I do not accept the interpretation of Maynard Smith and Sondhi, and see no reason why all their results should not be explained by cellular response to a simple linear gradient. I think the essence of the difference between us is found on page 135. Roberts' results to me are a beautiful example of how we are simply dealing with a universal positional information, and all that is happening is that the response of the cells to the positional value is changed by the mutant. There is no need to generate any pattern.

I think the crux of the issue is that I believe that what changes is competence, and that one never has to generate a pattern. This, incidentally, has an interesting corollary which is that it is as easy to generate by these means a complex pattern like Stars and Stripes as it is to do the French Flag. In both cases, the cells have merely to interpret their positional information.

144

Reply by C. H. Waddington

Reply by C. H. Waddington

I can see the logical attraction of Wolpert's idea that one can account for pattern formation by generating a system of specifying positional information and a code-book for interpreting it, and that therefore 'one never has to generate a pattern'. But I think Lewis is much too facile about how the cells 'have merely to interpret' their positional information. If one says that the positional information (or the prepattern, come to that) is the same in an antenna as in a leg, this tells one nothing about the difference in pattern of those organs ; the whole burden of explanation is left to be borne by the different codes of interpretation in the two cases. And it seems to me very difficult to think of a plausible chemical system which would result in a normal Drosophila tarsus forming intersegmental membranes at five places (say, points 10, 25, 45, 70, and 100 per cent from the tip), while a four-jointed tarsus formed them at four places (15, 40, 65, and 100 per cent). What sort of a funny system of multiple thresholds are you going to invent ? On the other hand, since we are dealing with a material structure, with physical properties and affected by stresses from growth, and so on, it is relatively easy to generate a pattern direct, as in my analogy with a congealing drop of wax.

Organizational principles for theoretical neurophysiology

Michael A. Arbib
University of Massachusetts

We start by suggesting the need for organizational principles to complement the study of molecular mechanisms in theoretical biology. We then consider a simple model of control of eye movement to suggest somatotopy as an ordering principle for the description of neural architecture. We stress that the neonate's brain is no *tabula rasa*, and briefly discuss the problems of theoretical embryology raised by the requisite neural specificity. We present a class of feedback schemes, each designed to find a transformation which will bring a pattern to standard form, and give as example Pitts and McCulloch's model of the superior colliculus as a controller of gaze. The Pitts–McCulloch model is then contrasted with the Braitenberg–Onesto model for the control of ballistic movement to suggest a general strategy for distributed control of movement. The decision-making of a frog in snapping at one of several flies is then contrasted with the process whereby a vertebrate is hypothesized to choose one of several gross modes of action. Finally, we summarize and extend the discussion by presenting eight organizational principles for theoretical neurophysiology.

THE 'TOP-DOWN' APPROACH

By now, many hitherto mysterious properties of cells—the basic 'components' of all organisms—have been explained in terms of biochemistry and molecular biology, and many papers document the power of such explanations. It would be foolish to try to belittle such achievements, but it would be equally mistaken to be so dazzled by their success as to believe that biochemistry alone can unravel all the knotty problems of biology, and that the development of new theoretical approaches is unnecessary.

Going 'up' from the basic biochemistry and physics must be complemented by going 'down' from overall functional questions, as in the division of labour in Computer and Information Science between the component expert (using solid-state physics in trying to push to the ultimate reduction of size, increase in speed of operation, and flexibility of function for devices which are then to be built into computers) and the computer scientist (who studies how to put together large-scale organizations in terms of components of stipulated function to get

146

some overall sophisticated behaviour). Similarly, in biology the component or cellular level is the meeting ground for two quite different approaches. To explain how cells 'work', and their capacities for interaction, is the task of the biophysicist and biochemist ; while explaining how to organize large collections of such components seems to require such approaches of Computer and Information Science as automata theory and computer simulation.

To claim this is not to claim that the right tools for a theoretical biology are readily at hand. One of the greatest pitfalls facing the mathematician, engineer, computer scientist, or physicist turning theoretical biologist is that too much of his education has involved his mastering long-polished mathematical techniques, and finding that wide classes of problems can be solved simply by 'plugging-in' these techniques. With this as background, it is all too easy to believe that he can solve the biologist's problems by 'plugging-in' these techniques to biological situations, little realizing that a great period of induction and experimentation (for even theorists must experiment, even if only with symbolic constructs) was required to match technique to problem. But theory *is* required in biology—as in any science where constructs become subtle enough to escape the domain of the immediately observable and where the depth of argument comes to exceed the usual grasp of common sense—and what remains to be determined is not whether there shall be theoretical biology, but rather what forms theoretical biology shall take. The theorist who can make a substantial contribution will probably be one who combines an intimate knowledge of the experimental data of some restricted problem in biology with a broad command of theoretical techniques, and uses the interaction between his reformulation and reconceptualization of the data and his reworking of the techniques to evolve genuinely new insights into that particular biological problem—only to find that those insights are valid elsewhere. There is no recipe for this, but in this paper we shall at least present some organizational principles that anatomical, physiological, behavioural, and simulation studies suggest may provide a useful framework for theoretical neurophysiology.

INTRODUCTION TO SOMATOTOPY

To describe a neural network, we must discern order in its complexity. Here we suggest somatotopy as an ordering principle of neural architecture. To see the theoretical implications of this suggestion, we start with a simple consideration of the control of eye movements.

Consider the problem of applying control theory to the fixing of gaze upon a

147

stationary or slowly-moving object. Many authors would note that two crucial parameters were involved—the present angle of gaze θ, and the desired angle of gaze, θ_d. They would then ask what function of the desired and actual gaze is computed to determine the rotational acceleration $\ddot{\theta}$ of the eye.

Such an approach has proved fruitful in analyzing behaviour of biological systems, but may be dangerously misleading when it comes to unravelling the details of neural circuitry, for it suggests that we view the brain in terms of a central executive which manipulates a few variables such as θ and θ_d to issue such directives as the current value of $\ddot{\theta}$.

However, θ_d is not immediately available to the brain, but is instead encoded in terms of peaks of activity in a whole layer of neurons—the rods and cones of the eye. Again, in the case of eye dynamics, $\ddot{\theta}$ cannot be effective as a single control signal for a rotary actuator but must rather control the opposed activities of at least one agonist–antagonist pair of muscles—and even here, two signals are not enough, for the contraction of each muscle, itself a population of muscle fibres, must result from the overall activity of a whole population of motoneurons

Thus, although it would be possible to design a robot with a 'brain' structured as a centralized $(\theta_d, \theta) \mapsto \ddot{\theta}$ converter it would require special preprocessors to 'funnel down' the whole input array of retinal activity to provide the single number θ_d. Indeed, this scheme might make sense in a robot whose task was to track single targets rather than interact with complex environments, and whose effector was a single rotary actuator for which $\ddot{\theta}$ was an appropriate control signal. But if the output must be played out upon a whole array of motoneurons, as in the biological case, so that $\ddot{\theta}$ would have to be fed into an elaborate pro-cessor to be 'parcelled out', then one begins to doubt the utility of the centralized processor. In the section, 'A distributed-computation model for the control of gaze', we shall recall a scheme, due to Pitts and McCulloch, whose beauty lies in the simplicity of its demonstration that—at least in the case under discussion—a centralized processor may be dispensed with, and all computation may be carried out in distributed fashion in the layer or layers between the input and output arrays. In the section, 'Redundancy of potential command', we shall outline a related model of frog visumotor activity.

Some data about the frog visual system may help make our point about the importance of somatotopy in neural design : Lettvin, Maturana, McCulloch, and Pitts [1] found that most ganglion cells of the frog's retina could be classified as being one of four types—such as 'moving spot detectors' and 'large mov-ing object detectors'. What we want to emphasize here is the way in which the information from the four types of detectors is distributed in the brain. Their axons terminate (among other places) in a brain structure called the *tectum*, with the terminations forming four separate layers, one atop each other, with the properties that (a) different layers correspond to different types of detector ; (b) each layer preserves the spatial relations between the original cells (that is, there exists a direction along the layer corresponding to moving across the retina) ; (c) terminations stacked above one another in the four layers come from ganglion cells with overlapping receptive fields.

In discussing somatotopy in the layers of such a distributed computer, the reader should take note that we shall use the word somatotopy in an extremely broad fashion. In the input pathways of the visual system, position encodes position in visual space relative to the eye ; in the auditory system it encodes frequency of some parts of stimulation ; and in the tactile system, position on the body. It is only in the last case that the term somatotopy [from the Greek *soma* (body) and *topos* (place)] is strictly appropriate, since it preserves information about place on the body as we move from receptors to the central nervous system—retinotopy and tonotopy may better connote the respective situations in the first two cases. Again, in output pathways, position in a layer may encode the location of the target of a movement.

It should be noted that such relationships between two layers may preserve rough spatial relationships (up and down *versus* across), without preserving relative sizes. For example, in the layer in the human brain which receives touch information from the body, the fingers occupy a larger area than the trunk, since the brain needs detailed sensory information from the fingers if it is to control fine manipulation.

As we move away from the periphery to layers of the brain far removed from any predominant commitment to sensory modality or particular mode of action, we can expect position in the layer to have little direct correlation with bodily position—yet we hypothesize that position in the layer will still encode a crucial parameter of the cell's function. What we are saying is that a useful way to structure the apparent chaos of many parts of the brain is to describe such parts in terms of interconnected layers, where position within the layer is a crucial indicator of the functional significance of a cell's activity, and where an analysis of one patch of such a layer may yield an understanding of the function of the layer as a whole. It is in this somewhat over-extended sense of a positional code that we shall speak of somatotopy even in layers far from the periphery.

We do not claim fanatically the universal truth of the statement 'the brain is a layered somatotopic computer'. Rather we use it as a convenient slogan to remind us that it is high time that somatotopy—so long an important property for anatomists and physiologists—played its full role in our theories of brain function. Even in structures which are not layered—the reticular formation may be one—the positions of neurons will play a role that we cannot neglect in modelling their contribution to the overall function of the structure.

NEURAL SPECIFICITY AND THEORETICAL EMBRYOLOGY

As our discussion of somatotopy in the above section has emphasized, there is a great deal of specificity in neural architecture. In this section we shall present further examples of such specificity (for a full review, *see* Jacobson [2]) and hint at some of the modelling problems which they raise for theoretical embryologists.

The stimulus-response view of behaviour held that if a creature, with adequate receptors and effectors, were put in some complicated environment and 'punished' when it did something 'wrong' and 'rewarded' when it did something 'right', then eventually the correct connections would be made to enable the organism to function effectively in that environment. Translating this into neural terms, many people thought the nervous system was completely 'plastic' : that all connections could be—and in fact were—moulded by experience. We owe to Paul Weiss, Roger Sperry, and other workers the knowledge that plasticity is *not* unlimited, and that in fact there is a great deal of neuronal specificity—that is, *genetics constrains many details of neuronal connections which cannot be changed by experience unless there exist specific brain structures to exploit that experience.*

A newborn baby has to be able to suck, to breathe, to excrete, and so on, but it cannot do many other things at birth, and has to be able to learn how to do them. This cannot happen unless it has appropriate structures to implement learning. In other words, neural plasticity—paradoxically—requires a constraining neural specificity to be fully effective. This point may seem obvious, but is so often lost sight of that it may pay to belabour it: think of tossing a coin repeatedly. Every time it comes up heads, spray it with Chanel No 5, and every time it comes up tails spray it with stale cabbage juice. It hardly seems profound to doubt that the coin will eventually tend to come up heads rather than tails, but it may be helpful to explicate the grounds for our doubt.

First, the coin does not have receptors which allow it to distinguish Chanel No 5 from cabbage juice. Secondly, even if it could distinguish them, it has no inbuilt criteria to determine which is preferable. Thirdly, even if it could tell which was preferable it has no mechanism whereby it could make use of that knowledge to change its behaviour. Thus in looking at the embryology of the nervous system we have to look for specificity, whether in the direct sense of determining networks which will mediate innate behaviour patterns, or to provide the adaptational substrate to enable the organism to adapt its evolutionary heritage to the exigencies of its own environment. We have to understand how appropriate receptor and effector arrays can be structured, how basic drive mechanisms can be 'built into' the organism so that it can shape its behaviour on the basis of some evolutionarily determined criteria of biological usefulness or destructiveness, and we must understand—at least in mammals—the determination of a sufficiently rich cortical structure to allow sophisticated learning.

To enhance the latter point by a striking contrast, we may recall Paul Weiss' intriguing experiments [3] in which the forelimbs of a salamander were reversed in the larval stage. When such a salamander grew to salamanderhood, whenever it would see some food in front of it, the brain would send the appropriate command of 'advance', but the neural circuitry in the brainstem which interpreted the command did not 'know' that the forelimbs were back-to-front and so would send the sequence of muscular activation which would cause the forelimbs to make the animal scurry away from its food. No matter how long the animal was exposed to this unfortunate situation it could never learn what was wrong—or, at least, if it learned what was wrong, it could not do anything about it. Thus we see the necessity for adequate structure if learning is to ensue.

Notice that what we are talking about in the nervous system is not the development of individual organs *per se*, but rather the development of functional systems which involve the whole organism. The animal at birth has to be able to take tactile stimuli on the lips and go through the 'computation' required to convert this into a sucking reflex. If we look at animals such as the guinea pig in which the hindlimbs are more important than the forelimbs at birth we will find the uneven development of the spinal cord which ensures that the hindlimbs are ready to function at birth. This is what the Russian physiologist Anokhin [4] refers to as *systemogenesis*—we have to think of the nervous system not in terms of anatomically defined lumps of tissue, but rather in terms of an interacting overlapping collection of systems for carrying out biologically important functions. Thus our task becomes even more complicated when we realize that it is not enough to look at one small part of the body or the nervous system and explain how it grows, but we have to explain the sort of synchrony which allows

Organizational principles

functioning systems of various kinds to be available at birth and at later stages of maturation. At the moment we have models at the simpler stage of studying morphogenesis of single organs—an apparently necessary way-station in the evolution of our models—before we can tackle the synchrony problems of systemogenesis. At the moment, we look at one organ in a system and try to explain what sort of cellular interaction can give rise to its shaping. We may hope that, later on, when we understand this, we will have the intellectual apparatus in place to combine together our models of several systems to understand what sort of overall mechanisms allow coordination of their development.

Having established, in the first section, the cellular level as an appropriate intermediate between the study of macromolecules and organelles by the bio-physicist and biochemist, and the study of organismic control by the computer and information scientist, and having now seen the interest of understanding embryological processes, let us briefly mention some of the mechanisms at the cellular level which shape the overall form of the organism, including that of its nervous system. (The reader will find an excellent overview of 'the forces that shape the embryo' in Trinkaus [5].)

One mechanism whereby a tissue may change its form is that of autonomous change in cell shape. For example it is now well known that various micro-structures may be synthesized within cells during characteristic changes of shape, and that their destruction impairs such changes. Thus cells seem able to elongate themselves by producing microtubules aligned parallel to the axis of elongation. Again, cells seem able to constrict a portion of themselves by producing micro-filaments which can then contract to provide the constriction by a sort of 'purse-string effect' [6]. Schroeder [7] has combined such mechanisms to provide an elegant model of neurulation—the process whereby a plate of cells on the back of the embryo is formed into a trough which then rolls up into a tube running the length of the embryo to disappear beneath the surface of the back to form the rudiments of the spinal cord and brain.

Another mechanism whereby a tissue may change its form involves the com-bined effects of cellular adhesiveness and cellular motility. Such a mechanism helps us understand situations in which the attachments of cells change over time, but where there seem to be important specificities in the ensuing pattern of cellular attachments. Gustafson and Wolpert [8] (for an exposition see also Wolpert and Gustafson [9]) have given a masterly analysis of cellular movement and contact in sea urchin morphogenesis. Ede and Agerbak [10] have been able to correlate changes in adhesiveness of cells (and the consequent change in their motility) in normal and *talpid*[3] mutant chick embryos with changes in the developing limb pattern in these embryos, while Ede and Law [11] have expressed this correlation in the specific form of a computer simulation of limb development.

While elegantly showing how changes in cell shape, motility, or adhesiveness can provide mechanisms for morphogenesis—both in the nervous system and elsewhere—the above schemes do not make explicit how a cell 'knows' what contribution it is to make in the overall pattern. It is for this reason that other workers have developed the idea of 'positional information'. Here, the line of argument runs 'If the cell is to change appropriately it must have information about its position within the embryo (and perhaps it will need to consult a clock, too)'. An early approach to such positional information was in gradient theory [12]—if a source of some metabolite were located at one end of the axis and a sink at the other, with a uniform gradient in between, then the concentration of metabolite in any cell would signal its position on the axis. Wolpert [13] has suggested ways in which such a model needs refinement and elaboration, and Goodwin and Cohen [14] have instantiated Wolpert's ideas in a model in which position is signalled by the phase differences between families of pulses propagating with different delays from cell to cell. By contrast, automata theorists have shown how cells may be formed into complex arrays without explicit 'addressing'. Rather, each cell is capable of a finite number of states, and at any time the cell changes state in a way dependent upon its previous state and that of its neighbours. For example, von Neumann [15] exhibited a self-reproducing array with tens of thousands of components, but the cells were only capable of 29 states, and so could not 'know where they were'. Arbib [16] has attempted to place this approach in a more biological context. The work of Apter [17] should also be mentioned here. Other authors have compared the change of state rules used by von Neumann and others to the rewriting rules employed by linguists to 'grow' a sentence from its grammatical description, and are now exploring the applicability of formal linguistics to theoretical embryology [18, 19].

In considering the specificity of cellular connections, we must not be misled by estimates that the amount of information in DNA is far less than that contained in the connections of the brain, which some have taken to imply that connections in the brain must be random. To see this, consider the following computer program which comprises four instructions :

1. Set *n* equal to zero.
2. Print out *n*.

Organizational principles

3. Replace n by $n+1$.

4. Return to the second instruction.

If you observe a computer executing this program, it will emit a stream of numbers which is endless—at least till you have exhausted the capacity of the computer. Arguments that a comparison of the number of DNA bases with the number of connections in the brain shows that the brain must be a random network are as naive as comparing the four instructions of the above program with the number of positive integers and concluding that the sequence of positive integers, since it has more than four elements, must be a random sequence !

From programming computers we know the flexibility of programs having loops within them which are hierarchically structured to provide for a great deal of economy in the way we specify processes. As a biological example of a plausible 'use' of such 'nested subroutines', we may cite the retina of the frog. We have already cited the structure of the retinal output, but turning to the circuitry within the retina, we may note that Lettvin and Maturana have schematized the connections between the interneurons of the retina's second layer and the ganglion cells as falling into two or three segregated layers. A plausible wiring scheme would then prescribe that certain types of axons from the interneurons terminated in one layer and so are highly likely to connect one level of the dendrites of the ganglion cells while other types of axons bearing different transforms of the visual input would terminate in the other layer thus hitting other parts of the ganglion cell dendrites. By this means, one can very simply specify how to get a retina that would function perfectly for the frog trying to snap flies in its world, without having to specify point-by-point interconnections. Hence, a sort of 'nested subroutine' approach could probably explain a great deal of the specificity of the nervous system without requiring an immense investment in genetic material. Such economy of genetic prescription augurs well for economy of functional description when we come to describe organizational principles for neurophysiological processes.

A DISTRIBUTED-COMPUTATION MODEL FOR THE CONTROL OF GAZE

Pitts and McCulloch [20] presented a feedback scheme designed to find a transformation T, from among a group G of possible transformations of patterns which are played, say, upon the retina, which will transform a pattern ϕ to a standard form $\phi_0 = T\phi$. (Arbib and Didday [21] consider the case in which we also use transformations of output activity to assure that the relation between input and output is in standard form.)

154

They generate the transformation in two steps :
(1) Associate with each pattern ϕ an 'error vector' $\mathbf{E}(\phi)$ such that $\mathbf{E}(\phi)=0$ if and only if ϕ is in standard form.
(2) Provide a scheme ω which will associate with each error vector a transformation which is error-reducing—that is, for all patterns ϕ we demand that $\mathbf{E}(\phi)$ be reduced after $\omega(\mathbf{E}(\phi))$ is applied to ϕ :

$$\|\mathbf{E}[\omega(\mathbf{E}(\phi)) \cdot \phi]\| \leqslant \|\mathbf{E}(\phi)\| \qquad (1)$$

with equality only in case $\mathbf{E}(\phi)=0$.

There are two main implementations of such a feedback scheme, only the second of which was considered by Pitts and McCulloch. [Henceforth, let us use W_ϕ to abbreviate $\omega(\mathbf{E}(\phi))$.]

In a *ballistic scheme*, ω is so structured as virtually to reduce the error to zero in one step :

$$\mathbf{E}[W_\phi \cdot \phi] \doteq 0 \text{ for all patterns } \phi.$$

A controller would then proceed as follows :
1. Given ϕ, compute $\mathbf{E}(\phi)$ and thus W_ϕ.
2. Form $W_\phi \cdot \phi = \hat{\phi}$.
3. Proceed on the assumption that $\hat{\phi}$ is sufficiently close to standard form.

Such a scheme is that used in ballistics where $\mathbf{E}(\phi)$ is the displacement of a bullet from its target, and W_ϕ is determined by the initial aim when the shot is fired—there is no possibility of making mid-course corrections. This is in distinction to a guided missile in which repeated corrections can be made.

In a *tracking scheme*, then, the error $\mathbf{E}[W_\phi \cdot \phi]$ may be little less than the previous error $\mathbf{E}[\phi]$—all we demand is that under a feedback scheme employing *repeated* application of the error-correction the error eventually goes to zero. A controller implementing tracking may proceed according to one of two schemes. The *implementation* mode corresponds to continually modifying the pattern until one is found which is in standard form ; the *planning* mode corresponds to continually modifying the transform until one is found which will bring the given pattern to standard form :
(1) Here ϕ will be the latest transformed version of the input pattern.
1. Replace ϕ by the new input pattern.
2. Use $\mathbf{E}(\phi)$ to obtain W_ϕ.
3. Form $W_\phi \cdot \phi$ to obtain the new pattern ϕ.
4. Is the new $\mathbf{E}(\phi)$ close enough to zero ?
   YES : Exit, ϕ may be treated as in standard form.
   NO : Go to 2.

(2) Here ϕ will be the *fixed* input pattern, and T will be the updated transform to be applied to ϕ.

1. Initialize T to be the identity transformation : $I\phi = \phi$.
2. Use $\mathbf{E}(T\phi)$ to obtain $\omega[\mathbf{E}(T\phi)]$.
3. Form $\omega[\mathbf{E}(T\phi)] \cdot T$ to obtain the new transform T.
4. Is the new $\mathbf{E}(T\phi)$ close enough to zero ?
   YES : Exit, $T\phi$ may be treated as in standard form.
   NO : Go to 2.

To guarantee that the schemes converge we need a stronger condition than (1)—one condition is that there exists some number δ such that $0 < \delta < 1$ and
$$\|\mathbf{E}(W_\phi \cdot \phi)\| \leqslant (1-\delta)\|\mathbf{E}(\phi)\| \text{ for all patterns } \phi \quad .$$
Convergence then follows from the fact that $(1-\delta)^n \to 0$ as $n \to \infty$.

There will be applications in which a controller may wish to use a mixed ballistic-tracking strategy—using a transform generator ω_1 to compute a first 'giant leap' to bring the pattern fairly close to standard form, then a second transform generator ω_2 to be used in a tracking strategy to iteratively 'fine tune' the pattern ever closer to standard form. In fact, this 'combined strategy' seems to be that employed in many biological systems [22].

The hard work in such a scheme is in actually defining an appropriate error measure E and then finding a mapping which can make use of error feedback to control the system properly so that it will eventually transform the input to standard form.

Let us see here how Pitts and McCulloch exemplified their general scheme in a plausible reflex arc from the eyes through the superior colliculus to the oculomotor nuclei to control the muscles which direct the gaze so as to bring the point of fixation to the centre of gravity of distribution of brightness of the visual input. (With our current knowledge of retinal 'preprocessing' we might now choose to substitute a term such as 'general contour information'—or any 'feature'—for 'brightness' in the above prescription. But that does not affect the model which follows.)

Julia Apter [23, 24] showed that each half of the visual field of the cat (seen through the nasal half of one eye and the temporal half of the other) maps topographically upon the contralateral colliculus. In addition to this 'sensory' map, she studied the 'motor' map by strychninizing a single point on the collicular surface and flashing a diffuse light on the retina and observing which point in the visual field was affixed by the resultant change in gaze. She found that these 'sensory' and 'motor' maps were almost identical.

M. A. Arbib

Pitts and McCulloch noted that excitation at a point of the left colliculus corresponds to excitation from the right half of the visual field, and so should induce movement of the eye to the right. Gaze will be centred when excitation from the left is exactly balanced by excitation from the right. Their model [20, Figure 6] is then so arranged, for example, that each motoneuron controlling muscle fibres in the left medial rectus and right lateral rectus muscles, which contract to move the left and right eyeballs, respectively, to the right should receive excitation summing the level of activity in a thin transverse strip of the left colliculus. This process provides all the excitation to the right lateral and medial left rectus, that is, the muscles turning the eye to the right. Reciprocal inhibition by axonal collaterals from the nuclei of the antagonist eye muscles, which are excited similarly by the other colliculus, serve to perform subtraction. The computation of the quasi-centre of gravity's vertical coordinate is done similarly. (Of course, computation may be performed by commisural fibres linking similar contralateral tectal points, instead of in the oculomotor nuclei.) Eye movement ceases when and only when the fixation point is the centre of gravity.

It must be emphasized that the reflex for which we have just summarized a crude, though instructive, model would be subject to 'higher control' in normal function. For example, 'interest' might be the criterion for determining which area of the visual field to examine, with the reflex determining the fixation point within the region (compare the fine tuning servo on a radio receiver) — gaze may then remain fixed at that point until it is 'adequately' perceived. Conversely, a sudden flash may usurp the averaging operations to dominate the reflex control of gaze momentarily, forcing the organism to attend at least briefly to a novel stimulus.

DISTRIBUTED MOTOR CONTROL

It should be noted that even if the mathematical equations formalizing the Pitts–McCulloch scheme for centering of gaze were to contain a damping term to prevent the eyeball from undergoing continual oscillations, it still has the defect of being essentially a tracking model, whereas the reflex 'snapping' of gaze toward a flash of light is essentially ballistic. In fact, human eye movements can be either ballistic or tracking. Typically, a human examining a scene will fixate on one point of the visual field then make a saccadic movement (the term for a ballistic eye movement) to fixate another point of the visual scene, until satisfied that he has scanned enough of the scene to perceive his current environment. However in other situations — such as watching a car go by before crossing the street — he will fixate upon an object and then track it. In man, various cortical

157

Organizational principles

areas can modulate activity in superior colliculus, and Bizzi has found in one of them—the so-called frontal eye field, which is in the frontal cortex—that there are three types of cells, type I which are active in ballistic eye movements, type II which are active in tracking eye movements, and other cells more concerned with head movements than with eye movements. Perhaps a similar situation will be found on closer examination of superior colliculus. In any case, it does seem that the Pitts–McCulloch scheme is more suited to the tracking mode than to the ballistic mode. To rectify this, let us then present another model, due to Braitenberg and Onesto [25], for a distributed computer controlling ballistic movement. (It should be mentioned that they conceived their model as a model of the cerebellum, but subsequent investigations have revealed so much new data about the cerebellum that their model cannot stand as a model of the cerebellar cortex without drastic modification. The reader may find a thorough critique of cerebellar modelling in Boylls and Arbib [26], but it would not seem fruitful to present the details here, for our aim in this paper is not to say 'Here is the correct model for the function of a certain subsystem of the brain', but rather to say 'Here is a fruitful way to go about modelling brain function'. In this spirit we present models which give one new principles of organization, hoping in this way to spur much further work to find the biological implementation of these principles in neural circuitry ; or to see their refinement in the design of control circuitry for robots.)

When a shot is fired from a gun two forces are involved—the explosion that propels the projectile towards the target, and the braking force that results when the projectile hits the target (if the target were to step aside, the projectile would not stop in the position at which it was originally aimed). Ballistic movements in animals also involve this 'bang-bang' control. There is an initial burst of acceleration as the agonist contracts and the antagonist muscle relaxes ; an intervening quiet period ; and then the final deceleration as the antagonist contracts. Experiments on rapid flexion and extension of joints have shown that muscle activation occupies only a small portion of the movement, that the duration of this activation does not seem to be related to the extent of the movement. Thus the duration of the movement seems to be determined mainly by the timing, relative to the 'go' signal, of the 'stop' signal (which has to be determined by the brain, rather than being imposed by the environment, as it was in our projectile example). Braitenberg and Onesto thus proposed a network for converting space into time (a subtle alchemy !) by providing that the position of an input (encoding the desired target position) would determine the time of the output (which would trigger the 'slamming on of the brakes'). The scheme has a linear array of output cells whose output circuitry is so arranged that the firing of any one of them will yield the antagonist burst that will brake the ballistic movement. There are two systems of input fibres each arranged in the same linear order, with position along the line corresponding to angle of flexion of the joint. The first class, which we shall call the C-fibres, connect to a single output cell. The second class, which we shall call the M-fibres, bifurcate into fibres which contact each cell in the array. The speed of propagation along these parallel fibres is such that the time required to go from one point in the array to another corresponds to the time the joint requires to move between the corresponding angles.

The controller then elicits a ballistic movement by firing three signals—one to trigger the agonist burst which will initiate movement, one on the C-fibre corresponding to the initial joint position, and one on the M-fibre corresponding to the target position. If we assume that an output cell can respond to parallel fibre input only if it has received C-fibre input, we see that only the output cell corresponding to the activated C-fibre will fire, and it is clear that its time of firing will correspond to its distance from the activated M-fibre. Thus it will elicit the braking effect of the antagonist burst at precisely the right time.

Note that, in the above scheme, we could relieve the controller of having to 'know' where the joint is, by having a feedback circuit continually monitor joint position and keep the appropriate C-fibre activated.

While we do not claim to have modelled *the* way the nervous system controls movement, what we have shown is that *a plausible subsystem for vertebrate nervous systems may be a distributed motor controller of a type in which position of the input on the control surface encodes the target to which the musculature will be sent*. Further, we might expect that—akin to the result of merging the Pitts–McCulloch scheme with the Braitenberg–Onesto scheme—if an array of points is activated on the input surface, the system will move to the position which is the 'centre of gravity' of the positions encoded by that array.

It should be noted that a full elaboration of this scheme would involve hierarchical arrangements. For example, in fixating a new point in space, increasing angles of deviation might require movement of eyes alone, then of eyes and head, and then of eyes, head, and trunk. Thus the output of the motor-computer would not control a single joint, but would control a whole hierarchy of subcontrollers, whose behaviour would of course be modified by the low-level postural controllers in the brainstem and spinal cord. We should also add that the scheme must be elaborated to provide for generating particular velocities, and so on. To

159

caricature it crudely, one may conjecture that such an option has evolved through the development of circuitry which can control tracking movements internally, rather than driving them through sensory channels.

REDUNDANCY OF POTENTIAL COMMAND
To exemplify this discussion, let us present a model of the reticular formation, and then relate it to a model of frog visuomotor behaviour which involves layered distributed computation. First, we need to comment on the idea of 'Redundancy of Potential Command'. If we take the position that perception of an object generally involves the gaining of access to 'programs' for controlling interaction with the object, rather than simply generating a 'label' for the object, we must emphasize the *gaining* of access to a program rather than the *execution* of the program—one may perceive something and yet still leave it alone. Thus in gaining access to the program, the system only gives it *potential* command, further processing being required to determine whether or not to act. A key question will thus be 'How is the central nervous system structured to allow coordinated action of the whole animal when different regions receive contradictory local information?' McCulloch suggested that the answer lay in the *Principle of Redundancy of Potential Command* which states, essentially, that command should pass to the region with the most important information. He cited the example of the behaviour of a World War I naval fleet controlled—at least temporarily—by the signals from whichever ship first sighted the enemy, the point being that this ship need not be the flagship, in which command normally resided.

McCulloch further suggested that this redundancy of potential command in vertebrates would find its clearest expression in the reticular formation of the brain stem (RF). Kilmer and McCulloch then made the following observations towards building a model of RF.

(1) They noted that at any one time an animal is in only one of some 20 or so gross *modes* of behaviour—sleeping, eating, grooming, mating, urinating, for example—and posited that the main role of the core of the RF (or at least the role they sought to model) was to commit the organism to one of these modes.

(2) They noted that anatomical data of the Scheibels [27] suggested that RF need not be modelled neuron by neuron, but could instead be considered as a stack of 'poker chips', each containing tens of thousands of neurons, and each with its own nexus of sensory information.

(3) They posited that each module ('poker chip') could decide which mode was most appropriate to its own nexus of information, and then asked, 'How can

the modules be coupled so that, in real-time, a consensus can be reached as to the mode appropriate to the overall sensory input, despite conflicting mode indications from local inputs to different modules ?'

This was the framework within which Kilmer, McCulloch, and Blum [28] designed and simulated the compartment model, called S-$RETIC$, of a system to compute mode changes, comprising a column of modules which differed only in their input array, and which were interconnected in a way suggested by RF anatomy.

Pitts and McCulloch's model (*see* the section, 'A distributed-computation model for the control of gaze') of the superior colliculus (which is the cat's 'equivalent' of the frog's tectum) was offered as a plausible explanation of how an animal might fixate its gaze at the 'average' or 'centre of gravity' of a field of illumination. For us, their scheme has the added significance that it showed how to design *a somatotopically organized network in which there is no 'executive neuron' which decrees which way the overall system behaves—rather the dynamics of the effectors, with assistance from neuronal interactions, extracts the output trajectory from a population of neurons, none of which has more than local information as to which way the system should behave.*

If we paraphrase our interpretation of the significance of the Pitts and McCulloch model of the superior colliculus to say that it showed how 'the organism can be committed to an overall action by a population of neurons none of which had global information as to which action is appropriate', we are struck by the similarity of the situation to that in our statement of the RF problem.

We may build on this to illuminate another system for the study of redundancy of potential command. The frog, which is normally immobile, will snap at any fly that comes into suitable range—'snapping' comprising a movement of the head (and, when necessary, the body) to aim at the fly, and the rapid extension of the tongue to 'zap' the fly. The situation seems very simple in that the frog does not seem to recognize flies as such—rather it will snap at any wiggling object, but will not snap at a stationary (that is, dead) fly. A frog confronted with two flies then presents us with a beautifully simple redundant command situation—normally the animal snaps at one of the flies, and so we have sought to model the brain mechanism that determines which of the flies will 'take command' of the frog. This could be explained in terms of a serial scan made of the tectum until a region is first found in which the activity in the four layers of ganglion cell termination in the tectum signals the presence of a fly—at which stage the scanner would issue a command to snap in the direction indicated by the current address of the scan. However, we argued that such serial processing is *not* a candidate for the frog's neural machinery because of the fact (among others) that the frog will sometimes snap midway between two flies—precisely the 'centre of gravity' effect one expects from an output system of the distributed computation type suggested by Pitts and McCulloch for centring of gaze. (Note that the above distinction between serial and distributed processing could not be made by asking *only* the usual question of sensory physiology, 'What information is relayed to the brain ?' but by also asking, 'How does the animal make use of such information to act ?')

However, we must note that while the Pitts–McCulloch model does yield integrated behaviour, it does not explain the 'usually-one-fly-effect'. Didday [29, 30] has offered a mechanism for this which, in retrospect, could be seen to bear a great resemblance to the Kilmer–McCulloch RF model. The observations on frog behaviour suggest three layers of processing, each involving distributed computation. The first layer operates upon the four layers of retinal information to provide for each region a measure of 'foodness'. The third layer does a modified Pitts–McCulloch type computation to direct motion of the frog to the position corresponding to the 'centre of gravity' of activity in the second layer. The task of the second layer is then very much akin to the task of the Kilmer–McCulloch RF. Where that model has an array of modules which must interact to get a majority favouring the same mode, the task of the second layer of Didday's hypothetical tectum is to turn down the activity of all but one region of (or from) the first layer. The essential mechanisms turn out to be very similar, and provide a plausible analogue for the 'sameness' and 'newness' neurons observed by Lettvin *et al.* [1]. The models differ in having all modes evaluated in each module, *versus* having a module identified with a mode. In any case, the study of frog behaviour sheds new insight on RF modelling, and suggests alternate hypotheses.

ORGANIZATIONAL PRINCIPLES

To place our discussion in perspective, let us make explicit the effect our views of brain structure will have upon our approach to modelling brain function, contrasting three possible strategies for theoretical neurophysiology with the principles that guide our own approach.

In certain invertebrates, we may find that the function of the system we wish to explain is mediated by a rather small neural network and so we might actually hope to track down, by explicitly simulating the behaviour of say 100 or so neurons, all the details of their interaction, and so obtain a plausibly complete explanation of how a locust, say, walks or flies. (See, for example, the beautiful review of 'Insect Walking' by Donald Wilson [31].)

When we turn to vertebrates, this strategy does not work, save in studies of peripheral circuits for muscle control, for there are just too many neurons. There are various strategies to take, depending on our ideas about structure, as to how one might make a model. The physicist has one ready answer for how we might model a system with millions or even billions of neurons. From his study of gases he would suggest statistical mechanics (or—in technical terms—'average the hell out of it'). Unfortunately, such averaging may destroy the very parameters of interest to us if we want to explain linguistic behaviour or coordinated motor behaviour as in a frog snapping at a fly. On the other hand, if we want to understand how the cooperative behaviour of many billions of neurons in the cortex gives rise to evoked potentials or electro-encephalograms, then some sort of statistical mechanical approach may well be worthwhile. However, a straightforward statistical approach to the very large system will not do for more detailed structural questions about complex information processing in brains.

In this context, it may be worthwhile to contrast two types of statistical model. Winograd and Cowan [32] were concerned with the fact that in as large a system as the brain one has both the likelihood of not specifying completely all neural interconnections accurately by genetic parameters and also the likelihood of many 'malfunctions' of components during actual information processing by the system. They wished to design networks with enough redundancy to ensure that the organism would not be too unreliable. Their strategy was to start from a very specific function they wanted a hypothetical 'nervous system' to undertake, and then provide ways in which they could transform this 'nervous system' into a new form which was sufficiently redundant that quite a lot of sloppiness in the 'wiring' and in the behaviour of the 'neurons' would still give correct overall function. This strategy of starting from a specific structure for computing some function and finding ways of introducing redundancy to make it resistant to certain types of damage both in growth and function is radically different from the strategy Cowan [33] took in his later work, in which he looked at interactions between thalamus and cortex only in terms of gross statistical parameters of their interconnection, and then asked if certain aspects such as cortical rhythms could be explained on this basis. In this case one wants only some crude parameters of overall system behaviour such as the period of rhythm recorded in gross potentials, and so one can 'average out' a lot of detail by statistical mechanical techniques. But if one wants to look at the detailed state-dependent

163

processing of inputs to get outputs then one has to impose far more structure, and study deterministic operation at a certain level.

Another approach to modelling a large system is that of compartment models. A brain modeller taking such an approach will not try to average over the complete system, but will look at the gross anatomy of the brain to subdivide the brain into various regions. He will thus try to simplify the problems of explaining one large region of the brain by breaking it down into a collection of interconnected 'black boxes' and see if by making multiple plausible guesses about those boxes and their interconnections he can put together a reasonably functional model of the overall system. It may then be easier to take those individual boxes with their plausible functions and try to model them back down to the cellular level than trying to do the whole thing directly. Perhaps one of the most interesting brain models of this kind is that of Kilmer, McCulloch and Blum [28] on the reticular formation. As we saw in the previous section, they used neuroanatomy to legitimize the compartmentation of the RF into a series of 'modules' ascending the longitudinal axis. Each module could then be modelled as a whole, and then the simulacra could be interconnected to get the overall change-of-mode behaviour which they posit to be exhibited by the reticular formation.

With this as background, I want to suggest eight principles which may help us understand how the human brain can control the complexities of a human's behaviour. (Their elaboration appears in my book *The Metaphorical Brain* [22].)
▶ 1. *Theory must be action-oriented*. One often talks as if human perception merely involved being able, when shown an object, to respond by naming it correctly. However, it is often more appropriate to say of an animal that it perceives its environment to the extent that it can interact appropriately with that environment. We can perceive a cat by naming it, true, but our perception may involve no conscious awareness of its being a cat *per se*, as when it jumps on our lap while we are reading and we simply classify it by the action we take as 'something-to-be-stroked' or 'something-to-be-pushed-off'. In computer jargon, then, we may say that perception of an object generally involves the gaining of access to 'programs' for controlling interaction with the object, rather than simply generating a 'label' for the object.
▶ 2. *Redundancy of potential command*. To repeat the argument of the previous section—you perceive something and yet may still leave it alone. Thus in gaining access to the program, the system only gives it *potential* command, further processing being required to determine whether or not to act. A key question will

thus be 'How is the central nervous system structured to allow coordinated action of the whole animal when different regions receive contradictory local information ?' In other words a brain must be able to 'Resolve Redundancy of Potential Command'.

We next state our third principle which, as we commented in the section 'Introduction to somatotopy', is known to all neuroanatomists but has been strangely neglected in brain theory.

▶ 3. *The brain 'is' a layered somatotopic computer.* In relating such a principle to statistical brain theories, I would suggest that we look at the brain in terms of specific structures, and use randomization only when we really feel we are ignorant. In other words, it is a justified strategy to make a model in which only certain parameters can be confidently specified—either by experiment or previous theory—and thus to set up random values for all the other parameters in the model. We may hopefully explain some of the functions for the system being modelled despite our gap of knowledge about structure, and then gain more detailed understanding of other functions as we find out more about further details of structure. This is something of an evolutionary strategy of biological modelling. What our third principle says is that a useful way to structure the apparent chaos of many parts of the brain is to describe such parts in terms of interconnected layers, where position within the layer is a crucial indicator of the functional significance of a cell's activity, and where an analysis of one patch of such a layer may yield an understanding of the function of the layer as a whole. We must add that muscles are not unitary devices with a single controller, but instead each muscle comprises a multitude of fibres controlled by a whole population of neurons. Since an *array* of input cells must thus control, with the aid of internal variables, an *array* of output cells, it seems at least plausible that it is an *array*, rather than a central executive, that intervenes. We thus have made the next hypothesis, encouraged by the results of the distributed computation and distributed motor-control models.

▶ 4. *The brain is a distributed, highly parallel computer, unlike most digital computers.* Our next point has been well introduced by our discussion above of neural specificity and theoretical embryology, which we may augment by suggesting that certain vagaries of brain structure only make sense when we see that the human brain is a variation on an evolutionary theme—and herein lies much of the power of *comparative studies* which use experiments on animal brains to help unravel the knot of human mentation. Similarly, we must note that the brain had to grow, rather than being wired-up by a technician who could

Organizational principles

refer item by item to a blueprint. Thus in explaining organismic behaviour we must stress that :

▶ 5. *Brains have evolved ; further, each brain has grown.* Evolution has given each animal a basic repertoire of skills for survival. The frog brain enables the frog to snap at flies ; the human brain enables the human newborn to suck and gaze and breathe and excrete. But in all animals to some extent, more so in mammals and perhaps exceptionally so in man, these evolutionary skills are augmented by individually acquired skills and memories. The social basis for much of human skills is, in fact, so great that nowadays man's adaptation of and to his environment is more dramatically a process of cultural than of biological evolution. These experiences, skills, memories cohere into what has been called an *internal model of the world*. This model is not a cardboard replica, but rather the memory structure that, for example, lets us walk into a strange room and, on the basis of visual stimuli from a brownish rhombus, know that a table is present and that we may put the papers we are carrying on that table without risk that they will fall to the floor. Thus :

▶ 6. *The brain must be able to correlate sensory data and actions in such a way as to build up an internal model of the world.* Of course it is not enough to perceive the presence of a table on which we place our papers if we then release our grasp of them some three feet from the table. Too many discussions of perception overlook our first point (action orientation) and so talk as if it were enough to classify an object. But we must know where it is if we are to interact with it :

▶ 7. *Perception is of where as well as of what.* Finally, we note that when walking around obstacles, we decide where to walk but, unless the ground is very uneven, need not concentrate on placing our footsteps ; further, it appears that the midbrain control of stepping can relegate to spinal mechanisms the maintenance of an upright posture as we step. These, and other, considerations (*see also* Greene [34-7]) suggest that it is useful to theorize that

▶ 8. *Both structurally and functionally, the brain is hierarchically organized.* As we saw in the previous section, the study of frog behaviour—which is very much in the spirit of these principles (though simplified by the non-adaptive aspect of the behaviour modelled)—sheds new insight on *RF* modelling, and suggests alternate hypotheses. Though the model is still but a crude oversimplification of the complexities of a real frog brain, we believe that our partial successes show that these organizational principles (or their evolved descendants !) will play a crucial role in future theoretical neurophysiology.

M. A. Arbib

Notes and References

This paper has grown out of brief remarks made at Serbelloni IV, and is based on two other conference papers : 'Organizational principles for embryological and neuro-physiological processes' presented at a conference on *Physical Principles of Neural and Organismic Behavior* 16-18 December 1970 at the Center for Theoretical Studies, University of Miami, Coral Gables, Florida ; and 'Transformations and somatotopy in perceiving systems' presented at the *Second International Joint Conference on Artificial Intelligence* held 1-3 September 1971 at Imperial College, London. The arguments of this paper are elaborated in the author's book *The Metaphorical Brain* (Wiley-Interscience : New York 1972). Preparation of this paper was supported in part by the Public Health Service under Grant No 2-R01 N S 09755-02 C O M from the National Institute of Neuro-logical Diseases and Stroke, and in part by the US Army Research Office Durham under Contract No D A H C 04-07-C-0043.

1. J. Y. Lettvin, H. R. Maturana, W. S. McCulloch and W. H. Pitts, What the frog's eye tells the frog's brain, *Proc. I R E, 47* (1959) 1950-9.
2. M. Jacobson, *Developmental Neurobiology* (Holt, Rinehart and Winston : New York 1970).
3. P. A. Weiss, Self-differentiation of the basic patterns of coordination, *Comp. psychol. Monogr. 17* (1941) 1-96.
4. P. K. Anokhin, Systemogenesis as a general regulator of brain development in 'The Developing Brain' (W. A. and H. E. Himwich, eds.) *Progress in Brain Research 9* pp. 54-86 (1964).
5. J. P. Trinkaus, *Cells Into Organs: The Forces that Shape the Embryo* (Prentice-Hall : Englewood-Cliffs, N.J. 1969).
6. P. C. Baker and T. E. Schroeder, Cytoplasmic filaments and morphogenetic movement in the amphibian neural tube, *Developmental Biology 15* (1967) 432-50.
7. T. E. Schroeder, Neurulation in xenopus laevis. An analysis and model based upon light and electron microscopy, *J. Embryol. exp. Morph. 23* (1970) 427-62.
8. T. Gustafson and L. Wolpert, Cellular movement and contact in sea urchin morpho-genesis, *Biological Reviews 42* (1967) 442-98.
9. L. Wolpert and T. Gustafson, Cell movement and cell contact in sea urchin morphogenesis, *Endeavour 26* No 98 (May 1967) 85-90.
10. D. A. Ede and G. S. Agerbak, Cell adhesion and movement in relation to the developing limb pattern in normal and *Talpid*[3] mutant chick embryos, *J. Embryol. exp. Morph. 20* (1968) 81-100.
11. D. A. Ede and J. T. Law, Computer simulation of vertebrate limb morphogenesis, *Nature 221* (1969) 244-8.
12. C. M. Child, *Patterns and Problems of Development* (University of Chicago Press 1941).
13. L. Wolpert, Positional information and the spatial pattern of cellular differentiation, *J. theoret. Biol. 25* (1969) 1-47.
14. B. C. Goodwin and M. H. Cohen, A phase-shift model for the spatial and temporal organization of developing systems, *J. theoret. Biol. 25* (1969) 49.
15. J. von Neumann (edited and completed by A. W. Burks) *Theory of Self-Reproducing Automata* (University of Illinois Press 1966).
16. M. A. Arbib, Automata theory and develop-ment, I, *J. theoret. Biol. 14* (1967) 131-56.
17. M. J. Apter, *Cybernetics and Development* (Pergamon Press : Oxford 1966).
18. A. Lindenmayer, Mathematical models for cellular interactions in development, *J. theoret. Biol. 18* (1968) 280-99, 300-15.
19. R. Laing, Formalisms for living systems, *Report 08226-8-T* (University of Michigan 1969).

Organizational principles

20. W. H. Pitts and W. S. McCulloch, How we know universals : the perception of auditory and visual forms, *Bull. math. Biophys. 9* (1947) 127-47.

21. M. A. Arbib and R. L. Didday, The organization of action-oriented memory for a perceiving system. Part I, the basic model, *Journal of Cybernetics 1* (1971) 3-18. (Part II, the hologram metaphor, by M. A. Arbib and P. Dev, will appear in a later issue.)

22. M. A. Arbib, *The Metaphorical Brain* (Wiley-Interscience : New York 1972).

23. J. Apter, The projection of the retina on the superior colliculus of cats, *J. Neurophysiol. 8* (1945) 123-34.

24. J. Apter, Eye movements following strychinization of the superior colliculus of cats, *J. Neurophysiol. 9* (1946) 73-85.

25. V. Braitenberg and N. Onesto, The cerebellar cortex as a timing organ, *Congress Inst. Medicina Cibernetica*, First Naples Atti., pp. 239-55 (1960).

26. C. C. Boylls and M. A. Arbib, Organizational principles in cerebellar modelling, *Progress in Biophysics* (to appear).

27. A. B. Scheibel and M. E. Scheibel, Structural substrates for integrative patterns in the brain stem reticular core, in (H. H. Jasper *et al.*, eds.) *Reticular Formation of the Brain* pp. 31-55 (Little, Brown : Boston 1958).

28. W. L. Kilmer, W. S. McCulloch and J. Blum, Some mechanisms for a theory of the reticular formation, in (M. Mesarovič, ed.) *Systems Theory and Biology* pp. 286-375 (Springer-Verlag : New York 1968).

29. R. L. Didday, Simulating distributed computation in nervous systems, *Int. J. of man-machine Studies 3* (1971) 99-126.

30. R. L. Didday, A possible decision-making role for the 'sameness' and 'newness' cells in the frog, submitted to *Brain Behav. and Evol.*

31. D. M. Wilson, Insect walking, *Ann. Rev. Entom. 11* (1966) 103-22.

32. S. Winograd and J. D. Cowan, *Reliable Computation in the Presence of Noise* (MIT Press : Cambridge, Mass. 1963).

33. J. D. Cowan, *Statistical Mechanics of Nervous Activity*. Technical Report J D C 69-1 (Committee on Mathematical Biology, University of Chicago 1969).

34. P. H. Greene, New problems in adaptive control, in (J. T. Tou and R. H. Wilcox, eds.) *Computer and Information Sciences* (Spartan : Washington 1964).

35. P. H. Greene, Models for perception and action, *Proceedings of the First Annual Princeton Conference on Information Sciences and Systems* pp. 245-53 (Dept of E E, Princeton University 1967). In relation to which see numerous articles by Russian workers in *Biophysics*, the English translation of the Russian *Biofizika*, in the section entitled 'Biophysics of Complex Systems. Mathematical Models'.

36. P. H. Greene, *A.* Essential features of purposive movements ; *B.* Cybernetic problems of sensorimotor structure : introductory remarks and survey of a study ; *C.* Seeking mathematical models of skilled actions ; *D.* An aspect of robot control and nervous control of skilled movements ; *E.* Coordination of two effectors and transfer of adaptation, *Institute for Computer Research*, Chicago, *Quarterly Progress Report No 16* (1968) Sections III A-D.

37. P. H. Greene, *The Theory of Tasks: Cybernetic Problems of Sensorimotor Structure.* Book in preparation (preliminary material may be found in references [35, 36]).

Stochastic models of neuroelectric activity

Jack D. Cowan
University of Chicago

Theories of how the brain works take many different forms. Some are essentially combinatorial, and are posed in terms of computer science. The hypothesis is made that certain neural nets function as pattern detectors, or as simple memory stores, for example. Model nerve nets are then devised to carry out such tasks in an efficient manner. This approach has been used recently in a very interesting way to provide a series of models for cerebral and cerebellar cortex, and for the hippocampus [1-3]. The end product is a series of predictions about the functional characteristics of nerve cells and synapses, and an interpretation of some of the anatomical details of neural nets. Models of this kind serve a very useful purpose, but they are difficult to test physiologically, since they do not generate predictions about neural dynamics, that is, about the spatio-temporally organized patterns of activity recorded in brains, which evidently underlie signal transmission. To serve this purpose, theories are required which emphasize the temporal aspects rather than the purely combinatorial aspects of the electrical activity of cells and populations. In this paper I wish to give a short account of some recent investigations, the intention of which is to provide a theory of this activity.

Histological studies of the brain show it to be a densely-packed net of some ten billion nerve cells, and countless satellite cells, the whole occupying a container of some two thousand cubic centimetres. Studies carried out some twenty years ago [4] led to the conclusion that any theory of neural nets must depend upon statistical considerations in which the connections of individual cells are not nearly so important as the global pattern of connectivity in populations. More recent investigations have led to a modification of this view [5, 6]. Nerve cell fibres do not make connections haphazardly, but appear to be directed to specific groups of cells. If the statistical view is to remain tenable, it must be that of a structured and heterogeneous population, perhaps subdivided into reasonably homogeneous populations, each of which is connected in a particular fashion to other homogeneous populations. Somewhat related conclusions about the statistical nature of neuroelectric activity can be derived from a study of the electrical responses of cells and populations. Only averages appear to be reliable indicators of neural activities. At the cellular level, characteristic discharge patterns

Stochastic models of neuroelectric activity

are enormously variable, especially in unanesthetized animals [7]. Similarly, population records show considerable variability in both the 'spontaneous' and 'evoked' cases [8]. There is thus an appearance of randomness in most if not all neural nets, both in the fine details of interconnection, and in the individual records of activity. Our problem is to give some expression to this randomness in developing a semi-statistical theory of neural activity. In this paper only the temporal consequences of such a theory will be considered ; very important considerations concerning neural activity which is both spatially and temporally organized will not be covered here.

▶ *Statistics of cellular activity.* Microelectrode records of the discharge patterns of single cells show that a majority of the cells fire continually at a low rate, with occasional *bursts* of activity, and that stimulation changes such patterns in a variety of ways. Histograms of the intervals between successive spikes show a wide variety of shapes, characteristic of various random point processes. If plotted on a long time-scale such histograms can be fitted by a random burst process [9]. Other cells fire in regular bursts, the frequency and amplitude of which are controlled by stimuli [10]. To construct a model which will do justice to such data we start with a simplified electrical model of the cell membrane which goes back almost to the beginning of the century [11]. This consists of a linear RC circuit responding to superimposed current pulses. The resulting voltage drives an amplitude-sensitive pulse generator, so that if the voltage exceeds a certain threshold, a pulse is generated which propagates to other cells, and also resets the membrane voltage (See Figures 1 and 2).

Various elaborations of this are possible, for example the intensity of incoming current pulses may be made a function of the membrane voltage, reflecting voltage-dependent conductance changes produced in the membrane by synaptic excitation. On the assumption that the incoming pulse patterns are essentially 'shot noise', the variation in membrane potential may be represented as a Markov process [12]. In the simplest case, when incoming pulses produce either unit positive jumps (excitation) or else unit negative jumps (inhibition), a 'birth-and-death' process results [13]. Let $p(v, t|v_0)$ be the transition probability density for membrane potential, $\tau = RC$ the time constant of the membrane, and $p(v'|v, t)$ the cumulative distribution of jump sizes. Then

$$\frac{\partial P}{\partial t} = \frac{\partial}{\partial v}\left[\frac{v}{\tau}P\right] - \lambda_e[P(v, t|v_0) - P(v-1, t|v_0)]$$

$$+ \lambda_i[P(v+1, t|v_0) - P(v, t|v_0)] \tag{1}$$

170

J. D. Cowan

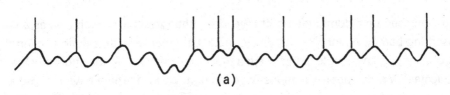

(a)

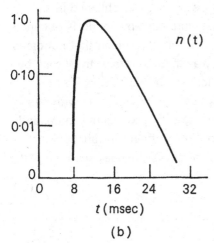

$n(t)$

t (msec)

(b)

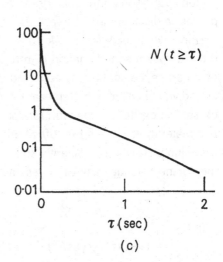

$N(t \geq \tau)$

τ (sec)

(c)

FIGURE 1

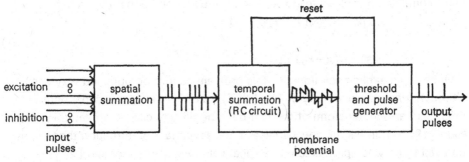

FIGURE 2

where λ_e and λ_i are the rate parameters of the incoming excitation and inhibition. At the threshold voltage $\theta(t)$, the process terminates and a spike is generated. The appropriate boundary condition is therefore

$$p(v'|\theta(t), t) = \begin{cases} 0 & v' \neq 0 \\ 1 & v' = 0 \end{cases}. \qquad (2)$$

It will be seen that the problem is a conventional one, that of a random walk to an absorbing barrier. The walk however consists of randomly occurring jumps

171

Stochastic models of neuroelectric activity

superimposed on a continuous drift away from the threshold, which acts as an absorbing barrier. As long as the time constant τ remains finite, solutions cannot be obtained in any simple fashion, even in the case where the threshold $\theta(t)$ is a constant. Various approximations have been introduced which have the effect of converting the continuous drift into a discontinuous one, thereby making the process a discrete random walk, the moments of which can be obtained in a reasonably compact form [14, 15]. A somewhat more tractable model is obtained if the diffusion approximation is introduced [16, 17]. It is assumed that hundreds of synapses are continually activated in such a way that the total activity seen by an individual cell may be represented as the difference of two shot-noise processes, an excitatory one with rate parameter λ_e and an inhibitory one with rate parameter λ_i. If ($\lambda_e - \lambda_i$) is sufficiently small, and λ_e and λ_i sufficiently large, and if the jump sizes are sufficiently small compared to $\theta(t)$, then the variation in membrane potential may be represented by the Ornstein–Uhlenbeck equation [18]

$$\frac{\partial P}{\partial T}=\frac{\partial}{\partial x}[xP]+\frac{\partial^2 P}{\partial x^2} \tag{3}$$

where

$$x=(v-\mu\tau)/\sqrt{(k\tau/2)}, \quad T=t/\tau \; .$$

μ is the drift parameter and k the variance parameter of the incoming Gaussian delta-correlated noise, white-noise. The boundary conditions now become:

$$P(-\infty, T|x_0)=P[\varepsilon(T), T|x_0]=0$$

and

$$P(x, 0|x_0)=\delta(x-x_0) \; . \tag{4}$$

Once again explicit solutions for this problem cannot be obtained very easily, although when the boundary $\varepsilon(T)=[\theta(T)-\mu\tau]/\sqrt{(k\tau/2)}$ is independent of time, the Laplace transform of the first passage density can be obtained in closed form [19, 20]. In certain special cases time-dependent perturbation theory can be used to provide approximate analytical solutions of the problem [21].

Let $w=x-\varepsilon(T)$, then eq. (3) becomes

$$\frac{\partial P}{\partial T}=\frac{\partial}{\partial w}[wP]+\frac{\partial^2 P}{\partial w^2}+f(T)\frac{\partial P}{\partial w} \tag{5}$$

subject to the conditions

$$P(-\infty, T|w_0)=P(0, T|w_0)=0$$
$$P(w, 0|w_0)=\delta(w-w_0) \tag{6}$$

where

$$f(T)=\dot{\varepsilon}(T)+\varepsilon(T) \tag{7}$$

172

Equation (5) represents a time-dependent perturbation of the Ornstein–Uhlenbeck equation, 'switched-on' at $T=0$, with boundary conditions on the half-line $(-\infty, 0)$. This will be recognized as a problem concerning quantum-mechanical harmonic oscillators. As is well known, one may write an integral equation equivalent to eq. (5) [22]. Let $P_0(w, T|w_0)$ be the Green's function solution of eqs. (5) and (6), when $f(T)=0$. Then :

$$P(w, T|w_0) = P_0(w, T|w_0)$$
$$+ \int_0^T \int_{-\infty}^0 d\tau dw' P_0(w, T-\tau|w') f(\tau) \frac{\partial}{\partial w'} P(w', \tau|w_0) \qquad (8)$$

which may be solved by iteration. The function of interest for the neural problem is the first-passage density

$$\mathscr{P}(0, T|w_0) = -\frac{\partial}{\partial T} \int_{-\infty}^0 dw' P(w', T|w_0) \quad . \qquad (9)$$

This is identical with the probability density of inter-spike intervals, provided the membrane potential resets to w_0, each time the threshold is reached. It can be shown [21] that the solution of eqs. (8) and (9) is :

$$\mathscr{P}(0, T|w_0) = \frac{|-w_0|}{\sqrt{2\pi}} e^{-T} (1-e^{-2T})^{-3/2} \exp\left[-\frac{w_0^2 e^{-2T}}{2(1-e^{-2T})}\right]$$

$$+ \left(-\frac{\partial}{\partial T}\right) e^{-T} (1-e^{-2T})^{-1/2} \exp\left[-\frac{w_0^2 e^{-2T}}{2(1-e^{-2T})}\right]$$

$$\times \sqrt{(2/\pi)} \int_0^T d\tau e^{\tau} f(\tau) \, \mathrm{erf}\left[-\frac{\beta}{\sqrt{(2\alpha)}}\right] + \dots \qquad (10)$$

where

$$\alpha = (e^{2(T-\tau)}-1)^{-1} + (1-e^{-2\tau})^{-1}, \quad \beta = w_0 e^{-\tau}(1-e^{-2\tau})^{-1} \quad .$$

It follows that if $f(T)$ is sufficiently small, an approximate solution is obtained of the solution of the Ornstein–Uhlenbeck process on the curved barrier $\varepsilon(T)$. Now

$$f(T) = \dot{\varepsilon}(T) + \varepsilon(T) = [\dot{\theta}(T) + \theta(T) - \mu\tau]/\sqrt{(k\tau/2)} \quad .$$

Thus $f(T)=0$ whenever $\theta(T) = \mu\tau$, or when $\theta(T) = \mu\tau + (\theta_0 - \mu\tau)e^{-T}$. In both cases $\mathscr{P}(0, T|w_0) = \mathscr{P}_0(0, T|w_0)$ [23-5]. A more general case is

$$\theta(T) = \theta_\infty + (\theta_0 - \theta_\infty)e^{-\delta T} \quad ,$$

equivalent to

$$f(T) = [(\theta_\infty - \mu\tau) + (\theta_0 - \theta_\infty)(1-\delta)e^{-\delta T}]/\sqrt{(k\tau/2)} \quad ,$$

where $0 \leqslant \delta \leqslant 1$. The threshold time-constant is about 40 msec compared to a membrane time-constant of about 5 msec, corresponding to a δ of 0·25. Evidently $\mathscr{P}(0, T|w_0) \sim \mathscr{P}_0(0, T|w_0)$ whenever $\sqrt{(k\tau/2)}$ is sufficiently large. For a cell receiving excitation from several hundred fibres under physiological conditions, this voltage is of the order of 2-3 mV, compared to an equilibrium threshold θ_∞ of about 10 mV. Thus when $\mu\tau \sim 7$ mV or more, $f(T)$ is at most 1. Computed results are comparable with experimental data. But of course such a model neglects many of the details of cellular morphology and physiology, but it does provide some insight into the nature of neural variability. The conclusions to be drawn from this study are that any unsynchronized excitation, or else intrinsic fluctuations in cell thresholds, can serve to produce the variability observed in cellular discharges (Figure 3).

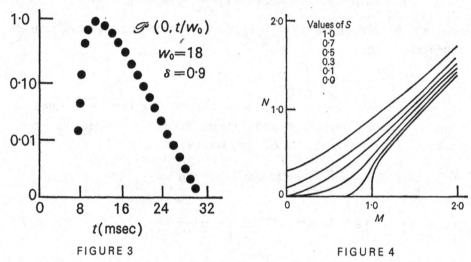

FIGURE 3   FIGURE 4

Given an approximation to the interspike interval density, one can use the Laplace transform to obtain moments of the density, and also to obtain such quantities as the *event density*, the conditional probability density that the membrane voltage reaches threshold at the instant t. The *conditional event density* is the sum of convolutions of the first passage density :

$$N(0, T|w_0) = \mathscr{P}(0, T|w_0) + \mathscr{P}(0, T|w_0) \otimes \mathscr{P}(0, T|w_0) + \ldots \qquad (11)$$

so that

$$N_L(0, S|w_0) = \mathscr{P}_L(0, S|w_0)[1 - \mathscr{P}_L(0, S|w_0)]^{-1} \quad . \qquad (12)$$

This expression is well known from renewal theory [26]. The instantaneous frequency is simply the reciprocal of the first moment of $\mathscr{P}(0, T|w_0)$, the mean interspike interval. Numerical calculations of this quantity indicate that there is a monotonic relationship between the intensity of incoming excitation, and the instantaneous frequency. Let $N = \tau n(w_0)$ be the average number of impulses within a time constant. Let $M = (\lambda_e - \lambda_i)\tau/\theta_\infty$ be the equilibrium voltage built

174

up in the membrane, that is, $\mu\tau$ in units of threshold. Let S be the r.m.s. noise voltage in threshold units, that is, $\sqrt{(k\tau)}/2\theta_\infty$. By curve fitting [27] (Fig. 4),

$$N = c \exp \left[\alpha(M-N) + \beta S^2 \right] \quad . \tag{13}$$

▶ *Statistics of neural net activity.* Given some insight into single-cell variability, it is natural to turn to a consideration of the activity in nets. The appropriate analysis is in terms of two-cell interactions. The simplest equation turns out to be a mixed Markov equation containing both drift and diffusion terms, and also jump terms containing time-varying parameters. Such an equation is not tractable and it is therefore necessary to develop a more simplified model of the neural response. Equation (13) provides the necessary starting point. It can be rewritten as

$$N_i = \exp \left[\sum_j^n \alpha_{ij} N_j - \delta_i \right] \tag{14}$$

for the ith (secondary) cell, receiving excitation from n (primary) cells, where the α_{ij} are defined in terms of synaptic conductances, membrane time constants, and neural thresholds [27]. If the cells also have *dead-times* or absolute refractory periods of duration r, before they become sensitive to excitation after the emission of an impulse, one obtains the equation :

$$N_i = (1 - rN_i) \exp \left[\sum_j^n \alpha_{ij} N_j - \delta \right] \tag{15}$$

Thus

$$N_i = L \left[\sum_j \alpha_{ij} N_j - \delta_i \right] \tag{16}$$

where $L[x]$ is the 'logistic' function of population growth, $e^x(1+re^x)^{-1}$. Alternately one has [28]

$$\ln \left[\frac{N_i}{1 - rN_i} \right] = \Sigma \alpha_{ij} N_j - \delta_i \quad . \tag{17}$$

To obtain an equation of motion from this equation it suffices to recall that there are conduction and synaptic *delays* of various kinds, all of which can be subsumed under a single time constant of the order of a few milliseconds at most. The result is an equation of the form

$$\left(\frac{d}{dT} + 1 \right) \ln \left[\frac{N_i}{1 - rN_i} \right] = \sum_j \alpha_{ij} N_j - \delta_i \tag{18}$$

where T is measured in units of the time-constant. If one adds an external input to each cell, and incorporates various constants into the terms, eq. (18) may be rewritten as

$$\frac{dx_i}{dT} + x_i(1-x_i) \ln \frac{x_i}{1-x_i} = \left(\varepsilon_i + \frac{1}{\beta_i} \sum_j \alpha_{ij} x_j \right) \cdot (x_i(1-x_i)) \tag{19}$$

where $x_i = rN_i$, and where ε_i contains the external input as a parametric excitation. In a certain sense this equation represents a generalization of the McCulloch–Pitts nerve cell model for individual pulse emission with an all-or-nothing threshold, to a model for pulse train emission with a continuous nonlinear characteristic.

It can be shown that for sufficiently large ε_i, the term $x_i(1-x_i)\ln[x_i/(1-x_i)]$ is small compared to $\varepsilon_i x_i(1-x_i)$, so that it can be neglected [29]. Equation (19) therefore reduces to :

$$\frac{dx_i}{dT} = \left(\varepsilon_i + \frac{1}{\beta_i}\sum_j \alpha_{ij}x_i\right)x_i(1-x_i) \tag{20}$$

where the coupling coefficient matrix $|\alpha|$ is as yet unspecified. To do this one must consider what would be an appropriate anatomy to incorporate. It has been argued elsewhere that eq. (15) might serve as a model of the interactions between neural nets in cerebral neocortex and in thalamus, and similar inter-actions in the cerebellum [28]. Large stellate cells in the thalamus are supposed to excite pyramidal cells in the neocortex. In turn these cells are supposed to inhibit the stellate cells. Both populations are supposedly embedded in a net of interneurons whose function is to maintain the appropriate synaptic coupling coefficients α_{ij} by some regulatory process. Finally there is an inhibitory input to the cortical pyramids and a corresponding excitatory input to the thalamic counter-parts. Such inputs may be external or they may be taken to be internal, in which case the thalamic cells may be said to act as *pacemakers*, driving the passive cells of the neocortex. The coupling coefficients are supposedly maintained so that

$$\alpha_{ij} + \alpha_{ji} = 0, \quad i \neq j \tag{21}$$

that is, all things being equal, the ith and the jth cells are in equilibrium. It also follows from eq. (20) that

$$\alpha_{ii} < 0 \quad . \tag{22}$$

It should be obvious that what has been set up is an analogue of the well-known Lotka–Volterra equations for predator–prey interactions between species, in which thalamic cells 'feed' predatory cortical cells. This topic has been ex-tensively investigated [30-3], and many of the recent results apply directly to the neural equations above, with trivial modifications.

Let q_i be the stationary states of the net, such that

$$\varepsilon_i\beta_i + \sum_j \alpha_{ij}q_i = 0 \quad . \tag{23}$$

176

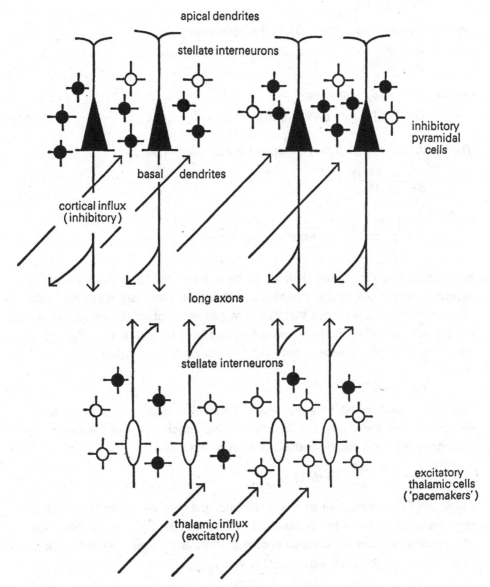

FIGURE 5
Thalamo-cortical circuits

Stochastic models of neuroelectric activity

The transformation

$$v_i = \ln \frac{x_i/q_i}{(1-x_i)}$$ (24)

may be used to reduce eq. (20) to the equivalent form

$$\frac{dv_i}{dT} = \sum_j^n \gamma_{ij} \frac{\partial G}{\partial v_j}$$ (25)

where $\gamma_{ij} = \alpha_{ij}/\beta_i\beta_j$, and where

$$G = \sum_j \beta_j [\ln(1+q_j e^{v_j}) - q_j v_j] \quad .$$ (26)

The function G is a Liapunov function for the system [34], for

$$\frac{dG}{dT} = \sum_i \frac{dv_i}{dT} \frac{\partial G}{\partial v_i} + \frac{\partial G}{\partial T}$$

$$= \sum_i \sum_j \alpha_{ij} q_i q_j \left[\frac{e^{v_i}}{1+q_i e^{v_i}} - 1 \right] \left[\frac{e^{v_j}}{1+q_j e^{v_j}} - 1 \right]$$

$$< 0$$

and G is evidently a positive definite function of v_i. Thus the net activity is asymptotically stable, and any oscillations about the stationary states gradually damp out. It is now assumed that there are intrinsic fluctuations associated with the activities of cells which can be represented by Gaussian delta-correlated noise, and that the variance of such diffusion terms is proportional to $-\alpha_{ii}$: that is,

$$k_i + \alpha_{ii} = 0 \quad .$$ (27)

This is not unreasonable ; the greater the feedback or dissipation intrinsic to neural activities, the greater the consequent fluctuations. One may therefore write a Langevin equation for the neural activities of the form :

$$dv_i = \sum_j \gamma_{ij} \frac{\partial G}{\partial v_i} dT + dB(T)$$ (28)

where $dB(T)$ is a Langevin term with variance parameter k_i. Such systems have been analyzed before in the Lotka–Volterra context [32, 33]. For the particular assumptions made here it is easily shown that the associated Fokker–Planck equation, for the probability density $f(v_1 v_2, \ldots, v_{2n}, T)$ is :

$$\frac{\partial f}{\partial T} = \sum_i \left[\frac{\partial}{\partial v_i} \left[\sum_j \gamma_{ij} \frac{\partial G}{\partial v_j} f \right] + \frac{k_i}{2} \frac{\partial^2 f}{\partial v_i^2} \right]$$ (29)

the equilibrium solution of which is :

$$f(v_1, v_2, \ldots, v_{2n}) = \exp[-\beta G] \quad .$$ (30)

178

J. D. Cowan

This is an example of the fluctuation-dissipation theorem, for although the system is non-linear, the dissipation is linear in the variable x_i. Moreover since G is a sum function, for each individual variable

$$f(v_i)dv_i \sim \exp[-\beta G_i]dv_i \qquad (31)$$

This result is extremely interesting because it provides some justification for the use of the Gibbs ensemble in the theory of neural nets. It is also possible to obtain the above equilibrium density in the following simple fashion. Suppose a single cell exists in a random environment, the effects of which are represented by Gaussian delta-correlated noise. Suppose that the cell is differentially sensitive to such fluctuations, in that it weights them by a factor proportional to its own activity. Let this factor be $x_i(1-x_i)$. Then the cell's activity may be represented by the nonlinear Langevin equation :

$$dx_i = -\beta x_i dT + \sqrt{(x_i(1-x_i)k)}d\xi_0(T) \qquad (32)$$

where $\xi_0(T)$ is standardized Gaussian delta-correlated noise, with zero mean and unit variance. Once again it is easily shown [35] that the equilibrium density for this process is given by eq. (31). As a matter of fact the transition density for this process is easily obtained and provides an approximate if rather crude means for estimating the first few moments of the transient response patterns of nerve cells.

Such results suggest a Gibbs ensemble treatment of neural nets. The main results have been presented elsewhere, in considerable detail [28], with a slightly different interpretation of the variable x. The equilibrium density takes the form

$$p(x_i)dx_i = \frac{x_i^{p-1}(1-x_i)^{q-1}dx_i}{B(p,q)} \qquad (33)$$

where $B(p,q)$ is Euler's β-function and where $p=\beta_i q_i/\theta$, $q=\beta_i(1-q_i)/\theta$. θ is the analogue of temperature in the net. It is a measure of the amplitude of fluctuation of activity and is given by

$$\theta = \beta_i \frac{\overline{(x_i-q_i)^2}/x_i(1-x_i)}{1-(x_i-q_i)^2/x_i(1-x_i)} . \qquad (34)$$

The smaller θ is, the more regular the activity, the larger θ is, the more 'burst-like' it is. The equilibrium density can be used to provide a variety of statistics relating to net activity. Among these are the 'interval' density (as seen through some kind of averaging), the amplitude density of membrane potential (seen through the same averager), the mean rate at which fluctuations in activity occur,

179

and certain correlation functions for the joint activities of coupled cells. Thus the interspike interval density is given by :

$$p(\tau_i)d\tau_i = \frac{\frac{1}{r}\left(\frac{r}{\tau_i}\right)^{p+1}\left(1-\frac{r}{\tau_i}\right)^{q-1}}{B(p,q)} \qquad (35)$$

The parameters p and q are themselves functions of the parameters q_i and θ, which reflect the cell's environment and its interaction with other cells. Some examples of this are shown in Figure 6. It will be seen that the distribution is approximately the Wiebull distribution characteristic of extreme-value statistics. The amplitude density of membrane potential can be obtained by way of the equilibrium equation

$$\ln\frac{x_i}{1-x_i} = \frac{\alpha}{\tau}v_i + \frac{\beta}{\tau} \qquad (36)$$

from which it follows that the variable

$$u_i = \exp\left[\frac{\alpha}{\tau}v_i + \frac{\beta}{\tau}\right]$$

has the density :

$$p(u_i)du_i = \frac{u_i^{p-1}}{(1+u_i)^{p+q}}du_i \quad . \qquad (37)$$

This will be recognized as a β-density of the second kind, the tails of which are of Pareto type [36], so that linear combinations of such random variables have amplitude densities, the tails of which are also of Pareto type. This is important for it permits the prediction of relationships between single unit activities and evoked membrane potentials and the electroencephalograms recorded by way of

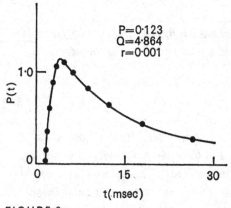

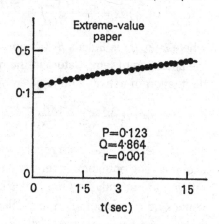

FIGURE 6

macroelectrodes (EEG), on the obvious assumption that to first order such potentials are the linear combination of cell membrane potentials.

Another statistical variable of some interest is the mean crossing rate $\overline{|\dot{x}_i|\delta(x_i-x)}$, the ensemble average of the rate of crossing of the amplitude x by the ith variable x_i. This expression can be derived exactly from the Gibbs theory. It is

$$\overline{|\dot{x}_i|\delta(x_i-x)}=2|\gamma_{ij}|x(1-x)p(x)q_j(1-q_j)p(q_j)\theta \tag{38}$$

in the case of two cells. More complicated expressions can be derived for larger nets. This expression provides a means for the estimation of γ_{ij}, the effective interaction coefficients of neural nets. Alternately,

$$\gamma_{ij}=\frac{1}{\theta}\overline{v_i\dot{v}_j} \tag{39}$$

so that γ_{ij} is proportional to the cross-correlation of the current driving the jth cell and the voltage built up in the ith cell.

As the final topic in this section, the effects of random parametric excitation on nets will be considered. It is assumed that the input coefficient ε_i has superimposed on it a Gaussian delta-correlated fluctuation, the effect of which is to produce random variations in the stationary states q_i. The equations of motion are as before [eq. (25)]. However

$$G=G_0+\sum_j \frac{\partial G}{\partial q_j}\delta q_j=G_0+G_1 \tag{40}$$

where G_0 is the unperturbed Liapunov function of eq. (26). Using known results for nonlinear Langevin equations [37], the following Fokker–Planck equation can be obtained for the transition density

$$f[v_1(T),v_2(T),\ldots,v_{2n}(T)|v_1(0),\ldots,v_{2n}(0)]=f(\bar{v}|\bar{v}_0) \ .$$

Let

$$\Omega_0=-\sum_i \frac{\partial}{\partial v_i}\left[\sum_j \gamma_{ij}\frac{\partial G_0}{\partial v_j}\right] \tag{41}$$

$$\Omega_1=-\sum_i \frac{\partial}{\partial v_i}\left[\sum_j \left(\frac{\gamma_{ij}}{\delta q_j}\right)\frac{\partial G_1}{\partial v_j}\right] \ . \tag{42}$$

Let $\bar{v}^0(T''-T')$ be the value of $\bar{v}$ at the instant $T'-T''$ before 0, due to the unperturbed motion, then

$$\frac{\partial f(\bar{v}|\bar{v}_0)}{\partial T}=\left(\Omega_0(\bar{v})+\int_{-\infty}^{T}\langle\Omega_1[\bar{v}(T)]\Omega_1[\bar{v}^0(T''-T)]\rangle_c dT\right)f(\bar{v}|\bar{v}_0) \tag{43}$$

where $\langle\ \rangle_c$ is the cumulant. It follows that

$$\frac{\partial f}{\partial T}=-\sum_i \left(\frac{\partial}{\partial v_i}\left[\sum_j \gamma_{ij}\frac{\partial G_0}{\partial v_j}f\right]+\sum_k \frac{\partial^2}{\partial v_i\partial v_k}\left[\sum_j \frac{\gamma_{ij}\gamma_{kj}}{\delta q_j^2}\left(\frac{\partial G_1}{\partial v_j}\right)^2 f\right]\right) \tag{44}$$

Stochastic models of neuroelectric activity

and using eqs. (26) and (40), that [38]:

$$\frac{\partial f}{\partial T} = -\sum_i \left(\frac{\partial}{\partial v_i} \left[\sum_j \gamma_{ij} \beta_j \left(\frac{q_j^0 e^{v_j}}{1 + q_j^0 e^{v_j}} - q_j^0 \right) f \right] \right.$$

$$\left. + \sum_k \frac{\partial^2}{\partial v_i \partial v_k} \left[\sum_j \gamma_{ij} \gamma_{kj} \beta_j^2 \left(\frac{e^{v_j}}{(1 + q_j^0 e^{v_j})^2} - 1 \right)^2 f \right] \right) \quad . \tag{45}$$

Given the equilibrium solution of this equation, the final phase space density is

$$\Phi(\bar{v}) = \int d\bar{v}_0 e^{-\beta G(\bar{v}_0)} f(\bar{v}|\bar{v}_0) \quad . \tag{46}$$

However the equilibrium solution to eq. (45) has not yet been found. A rough insight into what is likely to occur can be obtained from the one-dimensional nonlinear Langevin equation:

$$dx = x(1-x) d\xi_0(T) \tag{47}$$

the equilibrium transition density of which is

$$f(x|x_0) dx_0 = \left(\frac{x}{1-x} \right)^p \left(\frac{x_0}{1-x_0} \right)^{-p-1} dx_0 \quad . \tag{48}$$

This distribution is singular in x, tending to a delta function as $x \rightarrow 1$. The equilibrium density $\Phi(x)$ behaves in similar fashion. The conclusion to be drawn from this is that many cells coupled together will spend most of their time either completely off, or else firing at their maximum rates. This suggests that the effect of parametric excitation on the net is (paradoxically) to decrease somewhat the overall variability of the activity, and to produce more staccato, burst-like activity in the net.

▶ *Neural field theory.* The results of the above section indicate that a very particular set of assumptions is required to obtain what is essentially Hamiltonian dynamics from neural nets. In fact only the coupling relationships $\alpha_{ij} + \alpha_{ji} = 0$, $i \neq j$; $\alpha_{ii} = 0$ will serve. Any other law will produce either damped behaviour, for example, whenever $\alpha_{ii} < 0$, given the antisymmetry of α_{ij}, or else undamped behaviour whenever $\alpha_{ii} > 0$. The general conditions can be expressed in various ways, for example, by means of Liapunov theory [34]. The (locally) unstable cases are of great interest, for since the xs are bounded between 0 and 1, limit cycles are found in such cases. An example is seen in the system [39]:

$$\frac{dx_1}{dt} = (5 \cdot 2 + 14x_1 - 18x_2) x_1 (1 - x_1)$$

$$\frac{dx_2}{dt} = (-4 \cdot 8 + 18x_1 - 4x_2) x_2 (1 - x_2) \quad . \tag{49}$$

Such results suggest that it would be advantageous to set up a simple formalism which would permit the treatment of all these different cases, in a sufficiently

182

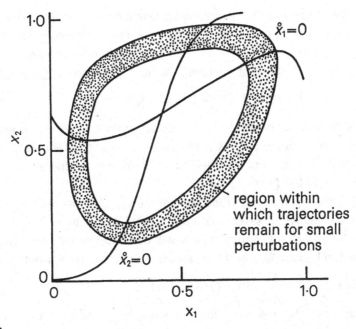

FIGURE 7

comprehensive fashion. The methods of non-equilibrium statistical mechanics could probably be used to some advantage, except that the sheer complexity of the details would probably vitiate the utility of such an approach.

From a theoretical point of view it seems reasonable to move from a detailed local description of cellular interactions to a more global one in which only the gross statistical features of connectivity are incorporated. Such statistical field theories have been investigated in the past [40-2]. In what follows a recent attempt will be sketched, the main intention of which is to incorporate both excitatory and inhibitory cell populations in a satisfactory manner [43]. Only the simplest problem will be considered in this paper, that of a homogeneous net of cells, each of which has a fixed threshold θ_i and a membrane response function $h(t)$. The cells are assumed to be randomly distributed with a volume or surface density ρ. Only one-dimensional problems will be considered here. Let $\xi(x)$ therefore be the mean number of fibres from cells in an infinite plane of unit thickness which synapse on a given cell at a distance x from the plane. $\xi(x)$ thus represents the global pattern of connectivity of the net. At any instant a proportion of the cells ρS are sensitive and respond to incoming excitation, and the remaining ρR cells have just fired, and are used or refractory. Let $F(x, t)$

183

Stochastic models of neuroelectric activity

represent the proportion of cells becoming active per unit time at the point s, at the instant t, that is, $F(x, t)$ is the rate at which cells become active. $F(x, t)$ will be called the 'activity' of the net, and in the field-theoretic description, such activity forms the 'sources' of the field. The field 'excitation' is given by :

$$\psi(x, t) = \int_{-\infty}^{\infty} F\left(X, t - \frac{|x-X|}{v}\right) \xi(x-X) dX \tag{50}$$

where v is the velocity of conduction of pulses in the net. $\psi(x, t)$ is evidently the rate of arrival of impulses to cells in the plane x, from all other cells in the net. Such excitation in turn generates new sources, so that in general

$$F(x, t) = \Psi[F(x, t-\tau), \psi(x, t-\tau)] \tag{51}$$

where $\Psi[\ ]$ is some nonlinear functional of the sources and excitation, and where τ is a delay which takes account of various retardations.

In what follows only temporal activity will be considered, so that the net will be assumed to be isotropic, and the excitation uniform over the net. Then

$$\psi(x, t) = F(t) \int_{-\infty}^{\infty} \xi(x-X) dX = kF(t) \ . \tag{52}$$

It is also assumed that the excitation varies slowly compared to the membrane time-constant, so that the mean integrated excitation built up in the cells is given by

$$\int_{0}^{t} \psi(x, T) h(t-T) dT = kF(t) \int_{0}^{t} h(t-T) dT \ .$$

$h(t)$ is taken to be a rectangular step of duration s and unit amplitude, whence

$$\int_{0}^{t} \psi(x, T) h(t-T) dT \sim skF(t) \ . \tag{53}$$

The duration s corresponds to the so-called period of latent addition of the cell. If s is sufficiently small, the proportion of cells which receive at least threshold excitation is substantially independent of whether or not the cells concerned have previously exceeded the threshold. Thus the activity at the instant t is proportional to the rate at which sensitive cells reach threshold at the instant $t-\tau$, and this rate equals the product of two terms, the proportion of cells which are sensitive at $t-\tau$, multiplied by the proportion of cells which receive at least threshold excitation at $t-\tau$. The first term is given by

$$1 - \int_{t-\tau-r}^{t-\tau} F(t') dt'$$

where r is the refractory period of the cells. The second term is essentially the event density (see the section, 'Statistics of cellular activity'). It follows from eq. (16) that the second term is

184

J. D. Cowan

$$L[skF]$$

so that eq. (51) takes the form :

$$F(t) = \left(1 - \int_{t-\tau-r}^{t-\tau} F(\lambda)d\lambda\right) L[skF(t-\tau)] \quad . \tag{54}$$

If the refractory period r is also of the order of s, then

$$F(t) = [1 - rF(t-\tau)]L[skF(t-\tau)]$$

or expanding about zero, assuming the retardation to be small,

$$\frac{\partial F}{\partial T} = -F + (1 - rF)L[skF] \tag{55}$$

where $T = t/\tau$.

If the net is assumed to consist of two distinct types of cells, excitatory and inhibitory [44], then eq. (55) gives rise to a pair of coupled equations, of the form :

$$\frac{\partial E}{\partial T} = -E + (1 - rE)L_e[s(k_1E - k_2I + P_1)]$$

$$\frac{\partial I}{\partial T} = -I + (1 - rI)L_i[s(k_3E - k_4I + P_2)] \tag{56}$$

where P_1 and P_2 represent external sources of excitation. These equations are evidently closely related to eq. (18) which was introduced to deal with the local aspects of net activities. Indeed starting from eq. (16) it would be equally appropriate to take as the basic equation

$$\left(\frac{d}{dT} + 1\right) N_i = L[\sum_j \alpha_{ij}N_j - \delta_i] \tag{57}$$

in which case the step from a net to a field representation of neural activity becomes obvious.

The dynamics of eq. (56) are of considerable interest, and are closely related to those of eqs. (57) and (18), especially when the refractory period r is small. The stationary states are easily obtained from solutions of the equations :

$$\ln \frac{E}{1 - (r+1)E} = sk_1E - sk_2I + sP_1$$

$$\ln \frac{I}{1 - (r+1)I} = sk_3E - sk_4I + sP_2 \quad . \tag{58}$$

These give rise to various types of singularities in the (E, I) plane, ranging from isolated to multiple stable foci, or else to isolated or multiple unstable foci, separated by limit cycles. Some examples are shown in Figure 8.

The transient behaviour of the function $E - I$ is of particular interest, since it is a measure of the net excitatory activity in the field. This activity ranges from

185

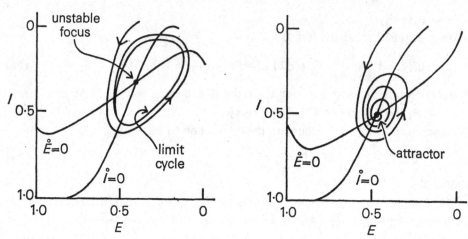

FIGURE 8

rapidly damped oscillations to maintained limit cycle oscillations, the particular form being set by the parameters k_i and P_i. Such a function is of particular interest because it provides a model for the generation of spontaneous and evoked potentials. Another important property of these equations is that the frequencies found in limit-cycle oscillations are monotonically related to the strengths of incoming excitation P_i, in much the same way as is the rate of pulse emission in individual nerve fibres [45]. There is thus coding of stimulus intensity into maintained frequencies of burst activity within the underlying net. The difference between this kind of coding, and the burst-coding obtained in the previous section from the Hamiltonian systems, is of course that the limit cycle activity is stable, and small perturbations of the parameters will not change the dynamics, except at certain boundaries. Equations (56) are therefore 'structurally stable' [46], and the stability results from the existence of dissipative rather than conservative dynamics. Structurally stable systems are surely more relevant to neural net theory than conservative structurally unstable systems, although it does not follow that the use of the canonical density for averaging is precluded, since, as has been indicated, such densities can be obtained from slightly dissipative noisy systems.

As a final point, the connection between the nonlinear field theory which has been outlined and such consequences of neural activity as perception, learning, and memory will be considered. As has been noted, the structural properties of the field dynamics are all highly dependent on parameters, and although the field is structurally stable with respect to small perturbations of the parameters, there

186

J.D.Cowan

are evidently boundaries in the parameter-space which separate the different dynamical possibilities. Any perturbation which takes the system across such boundaries will switch the system from one dynamical regime to another. Such changes are now called 'catastrophes'. It follows that if the parameters of the field are modifiable, then in principle long-lasting changes can give rise to catastrophes, and so one might expect to observe the sudden emergence or quenching of maintained oscillatory activity in the field, as a result of long-lasting modifications of synaptic connectivity. Now the outcome of the combinatorial approach to neural modelling is a set of predictions concerning which cells are excitatory or inhibitory, and which synapses are modifiable, as well as some account of the overall level of activity to be expected in certain nets, but not a detailed set of predictions about the specific kinds of changes to be found in the overall patterns of activity. Nonlinear field theory which incorporates modifiable parameters, and some of the known details of neuroanatomy, appears to be rich enough to provide a method by which the global effects of local changes in the nervous system can be understood, at least in simple terms.

▶ *Abstract.* A short account is given of some recent investigations concerning the dynamics of nerve cell nets. Various stochastic models of single cell activity are described, in particular the model which leads to the Ornstein–Uhlenbeck process on a curved absorbing boundary. The application of equilibrium statistical mechanics and of Fokker–Planck techniques to neural net models is then described. The paper concludes with a discussion of a recent attempt to model non-local neural interactions by means of a nonlinear field theory.

This work was supported in part by the Alfred P. Sloane Foundation and the Otho S.A.Sprague Memorial Institute.

References

1. D. Marr, *J. Physiol. 202* (1969) 437-70.
2. D. Marr, *Proc. roy. Soc. B 176* (1970) 161-234.
3. D. Marr, *Phil. Trans. roy. Soc. B* (in press).
4. D. A. Sholl, *The Organization of Cerebral Cortex* (Methuen : London 1953).
5. R. M. Gaze, *The Formation of Nerve Connections* (Academic Press : New York 1970).
6. M. Jacobson, *Developmental Neurobiology* (Holt, Rinehart and Winston : New York 1970).
7. G. L. Gerstein and N. Y.-S. Kiang, *Biophysical J. 1* (1960) 15-28.
8. M. A. Brazier, *Acta. Physiol. Neerlandica 6* (1957) 692.
9. D. R. Smith and G. K. Smith, *Biophysical J. 5* (1965) 1.

10. G.F. Poggio and L.J. Viernstein, *J. Neurophysiol. 6* (1964).

11. L. Lapique, *L'Excitabilité en Fonction du Temps* (Hermann : Paris 1926).

12. C.E. Molnar, *Thesis* (Massachusetts Institute of Technology 1966).

13. R.B. Stein, *Biophysical J. 5* (1965) 173.

14. M. ten Hoopen, *Biophysical J. 6* (1960) 435.

15. N.S. Goel, N. Richter-Dyn and J.R. Clay *J. theoret. Biol. 34* (1972) 155.

16. S.E. Fienberg, *Kybernetik 6* (1970) 227-9.

17. L.M. Ricciardi and R. Capocelli, *Kybernetik 8* (1971) 214.

18. G.E. Uhlenbeck and L.S. Ornstein, *Phys. Rev. 36* (1930) 823.

19. A.J.F. Siegert, *Phys. Rev. 81* (1951) 617.

20. B.K. Roy and D.R. Smith, *Bull. math. Biophys. 31* (1969) 341.

21. J.D. Cowan (in preparation).

22. R.P. Feynman and A.R. Hibbs, *Quantum Mechanics and Path Integrals* (McGraw-Hill : New York 1965).

23. M.C. Wang and G.E. Uhlenbeck, *Rev. mod. Phys. 17* (1945) 323.

24. W.M. Siebert, *Quart. Progr. Rept., M.I.T., R.L.E. 94* (1969) 281.

25. H. Sugiyama, G.P. Moore and D.H. Perkel, *Math. Biosci. 8* (1970) 323-41.

26. D.R. Cox, *Renewal Theory* (Methuen : London 1962).

27. P.I.M. Johannesma, *Thesis* (University of Nijmegen 1969).

28. J.D. Cowan, in (M. Gerstenhaber, ed.) *Some Mathematical Questions in Biology 2* (Amer. Math. Soc. : Providence, Rhode Island, 1970).

29. J.D. Cowan and E.H. Kerner (unpublished).

30. V. Volterra, *Leçons sur la Theorie Mathematique de la Lutte pour la Vie* (Gauthier-Villars : Paris 1931).

31. E.H. Kerner, *Bull. math. Biophys. 19* (1957) 121 : *21* (1959) 217 ; *23* (1961) 141.

32. E. Leigh, in (M. Gerstenhaber, ed.) *Some Mathematical Questions in Biology 1* (Amer. Math. Soc. : Providence, Rhode Island, 1968).

33. N.S. Goel, S.C. Maitra and E.W. Montroll, *Rev. mod. Phys.* 43 (1971) 231.

34. J.P. LaSalle and S. Lefshetz, *Stability by Liapunov's Direct Method* (Academic Press : New York 1965).

35. J.D. Cowan (unpublished).

36. W. Feller, *Introduction to Probability Theory and its Applications II* (Wiley : New York 1966).

37. R. Kubo, *Lecture Notes on Statistical Mechanics* (University of Chicago 1963).

38. J.D. Cowan and T. Sinnema (unpublished).

39. J.D. Aplevich, *Thesis* (University of London 1967).

40. R.L. Beurle, *Phil. Trans. roy. Soc. B 240* (1956) 669.

41. J.S. Griffith, *Bull. math. Biophys. 25* (1963) 111-20: *27* (1965) 187-95.

42. M. ten Hoopen, in (E.R. Caianiello, ed.) *Cybernetics of Neural Processes* (CNR : Rome 1965).

43. H. Wilson and J.D. Cowan, *Biophys. J. 12* (1972) 1.

44. J.C. Eccles *The Physiology of Nerve Cells* (Johns Hopkins Press : Baltimore 1957).

45. R.B. Stein, *Proc. roy. Soc. B 167* (1967) 64-86.

46. R. Thom, *Topology 8* (1969) 313-35.

Statistical and hierarchical aspects of biological organization

Michael Conrad
University of Miami

Biological nature has a compartmental structure. For example, consider an eco-system in the most general sense—as the aggregate of all the components of a complete biological system. The ecosystem could be thought of in terms of the configuration of certain fundamental components—say small molecules. How-ever this would obscure the organization of the ecosystem. In fact, biologists have found it useful to describe the ecosystem as a configuration of species, species as configurations of organisms, organisms as configurations of cells, and so forth.

The compartments of a biological system, except for single molecules, are open to materials which cycle through the ecosystem. Compartments are also open to energy, which is converted to lower-grade energy. There are many alternative pathways by which high-grade energy may be converted to lower-grade energy. It is these alternatives which provide the basis for information processing in biology, in particular the processes of reproduction. Self-reproducing systems must have a description or genotype, as well as decoding machinery ; the de-scription codes for the sequence of amino acids in proteins. These proteins, in turn, select the pattern of pathways which characterize the system.

Whether or not a compartment actually reproduces depends on its physical and biological environment—on the conformation of the ecosystem. In order to re-produce a compartment must be specialized, or adapted to its environment. The compartment must also be adaptable, or capable of coping with an uncertain environment. Adaptability is related to the ensemble of possible configurations of a compartment, or its possible modes of behavior. The adaptability and adapted-ness of a compartment may interfere with one another, in the sense that an increase in the ensemble of configurations entails a decrease in the specific relationship between the system and any given environment. Furthermore, the relations between compartments at different levels of organization may result in cross-level interference effects.

Each compartment of a biological system is more or less adaptable, in the sense of being associated with larger or smaller ensembles of configurations. For example, the gene pool may be more or less diverse, developmental patterns

189

Biological organization

more or less plastic, control mechanisms at the cellular, organismic, and societal levels more or less homeostatic, behavior patterns at the neural level more or less instinctive.

In this paper I would like to consider some of the factors which determine the way in which these statistical or control characteristics are distributed to the compartments of a biological system. In particular, I will consider a relation which ties compartments together in the sense of imposing certain restrictions on this distribution. This relation connects the control, computation, coding, and variability characteristics of biological systems to the noise characteristics of the environment, at least to the extent of expressing tendencies which might be expected in the course of evolution.

The approach which will be followed is along the lines of information theory and is based on the work of Shannon [1] and Ashby [2]. I will begin by considering the structure of biological organization. Then I will define statistical quantities which can be associated with adaptability and specialization, and will consider the specialization of labor in biological systems. This will provide a basis for discussing the relation between the statistical characteristics of a biological system and its environment, as well as the distribution of these characteristics to different compartments. These relations will be expressed in the form of a vector representation of biological organization. In the final sections I will develop a simple model of biological organization and will discuss (in terms of the model) some of the facts which have motivated the present approach.

▶ *Compartmental structure of biological organization.* Let $_k^h c_i^j$ represent compartment i at level j of compartment k at level h. For example, some cell could be compartment i of organism k. In general indices will be suppressed when they are not needed.

Compartments must be chosen so that they are non-intersecting as far as subcompartments are concerned. There are many classification schemes which fulfil this criterion. However, only certain units of organization are interesting from the biological point of view. Compartments are generally chosen in such a way as to minimize the importance of extrinsic relations—that is, on the basis of their relative degree of isolation.

A possible classification is listed in Table 1. The main levels include cells, organisms, species, as well as the ecosystem at the top level. In practice it is difficult to define these units precisely. However, each cell has a description, or genotype. The organism is a society of cells originating from a single cell. The genotype of this cell is the genotype of the organism. The species is a lineage of

organisms which share a common gene pool. The top level of the ecosystem
is the configuration of species. Each of these compartments may have certain
specialized subregions or organs : trophic structure in the ecosystem, including
grazing and detritus pathways ; social organization within the species ; organs
proper in the organisms ; organelles, including the genome, in cells.

TABLE 1

Possible classification of biological organization

Level (j)	Compartment
8	ecosystem
7	trophic level
6	species
5	society
4	organism
3	organ
2	cell
1	organelle or large molecule
0	small molecule

The classification is not entirely satisfactory since it does not eliminate certain
important relations between compartments. For example, the inclusion of species
precludes the inclusion of social organization involving symbiosis. These relations
must be included in the description of the ecosystem at the top level. Actually
the ecosystem consists of loosely coupled subecosystems, or communities.

▶ *States of compartments*. Each compartment has states. The states are configurations of subcompartments at some lower level. For example, the organism
has states—these are configurations of organs ; organs have states—these are
configurations of cells, and so forth. In general the state of compartment c_i^j is
determined by the configuration of the $_i^j c_k^h$, the subcompartments at level h.
Unless otherwise specified the state of a compartment will be determined by the
configuration of compartments at the next lower level ($h = j - 1$). The set of
states associated with each compartment will be denoted by capital letters and
the individual states by small letters. For example, the set of system states is
represented by $Z(z_\alpha \varepsilon Z)$ and the set of environmental states by $Z^*(z_\alpha^* \varepsilon Z^*)$. The
set of states associated with compartment c_i^j is Z_i^j, with $z_{i_\beta}^j \varepsilon Z_i^j$. The state set of all
compartments of the system other than c_i^j will be denoted by $\bar{Z}_i^j$, with $\bar{z}_{i_\gamma}^j \varepsilon \bar{Z}_i^j$.
Notice that we can identify z_α with the pair $(z_{i_\beta}^j, \bar{z}_{i_\gamma}^j)$ for some α. Distinct configurations may be identical from the macroscopic point of view. In this case they
are labeled by the same state.

Biological organization

Suppose that Z_i^j are states of interest, associated with some compartment of interest. The compartment is exposed to certain environmental states or external inputs, represented by Z^*. The compartment may also be influenced by complementary states or controlling inputs from other compartments, represented by $\bar{Z}_i^j$. The complementary compartments of a system are just those which are not included in the description of the compartment of interest, whether or not they are at the same level of organization. Of course the decomposition of a system into complementary compartments and compartments of interest is a matter of viewpoint. Environmental states are those over which a system has very little influence, at least in the short run.

The trajectory of a compartment is its sequence of states. A particular trajectory can be represented by the map

$$B : Z^*(t) X [Z_i^j(t), \bar{Z}_i^j(t)] \rightarrow [Z_i^j(t+\tau), \bar{Z}_i^j(t+\tau)] \tag{1}$$

where $(Z_i^j, \bar{Z}_i^j) = Z$, t is the time, and τ is some time interval. This is equivalent to the matrix

$$
\begin{array}{c|ccccccc}
 & z_1(t) & . & . & . & . & . & z_m(t) \\
\hline
z_1^*(t) & z_{11}(t+\tau) & . & . & . & . & . & z_{m1}(t+\tau) \\
. & . & & & & & & . \\
. & . & & & & & & . \\
. & . & & & & & & . \\
. & . & & & & & & . \\
. & . & & & & & & . \\
z_n^*(t) & z_{1n}(t+\tau) & . & . & . & . & . & z_{mn}(t+\tau)
\end{array}
\tag{2}
$$

where $z_{uv} = z_\alpha$ for some α. The set of possible trajectories is given by the set of maps $\{B\}$, where this is defined by the set of transition probabilities

$$\{f[z_\alpha(t+\tau)|z_\gamma^*(t), z_\beta(t)]\}$$

or their inverses $\{f[z_\beta(t)|z_\gamma^*(t), z_\alpha(t+\tau)]\}$.

The z_α consist of component states, for example, complementary states and states of interest. The complete state may give more information about the prior state and environment than the state of interest alone. In this case the complementary states act as a memory. The states of interest may also influence the complementary states. Such feedback makes it possible for states of interest to be controlled to a high degree of accuracy, or to change in a way which is independent of the environment. It is important that states can influence states at their own level as well as states at other levels.

The complete states may not be uniquely determined by the prior states and states of the environment. For example, the state of the system and environment may only determine the probability of the next state, that is, the transition proba-

M. Conrad

bilities are low. Alternatively, states may repeat in the columns or rows of the matrix [eq. (2)]. In this case the system has an equivalence class of histories since the inverse transition probabilities are low. Such selective loss of information about the past is an essential requirement for machine behavior [3, 4]. This is evident from the fact that most computations cannot be performed without reset operations. In general any finite automaton can be decomposed into simple reversible and simple irreversible machines, or actually in terms of the groups and semi-groups associated with these [5].

A purely mechanical system is reversible, since a complete knowledge of a small part of its trajectory determines the past and future uniquely. Thus, loss of information about environmental states must be associated with loss of detailed information about the states of the system, assuming that the environmental states are really part of the initial conditions. This is unavoidable if distinct microscopic states are indistinguishable, or groupable into classes, from the macroscopic point of view. Accordingly, the processes of computation are dissipative, or thermodynamically irreversible [6].

▸ *Diversity of states.* In this section I will discuss a quantity which expresses the uncertainty associated with an ensemble of states, or the diversity of states and their homogeneity of distribution. It is important to distinguish between such uncertainty, which amounts to variability of the compartment, and the diversity and homogeneity of a compartment in terms of its subcompartments. The composition of a compartment might be uncertain in this respect without being variable.

Consider an ensemble of compartments of type c_i^j. Uncertainty is measured by the entropy

$$\Gamma[Z_i^j(t)] = -\sum_{\beta} f[z_{i_\beta}^j(t)] \log f[z_{i_\beta}^j(t)] \tag{3}$$

where $\Gamma[Z_i^j(t)]$ is the state entropy of a representative compartment c_i^j and $f[z_i^j(t)]$ is the frequency of states z_i^j at time t. The state entropy increases with increase in the number of states (diversity) and equalization of the frequency of states (homogeneity). If the states are not discrete the phase space of the compartment must be divided into cells, and the entropy determined from the frequency with which these cells are occupied. The choice of cell size affects the value of the entropy only through a constant term.

The uncertainty of the system over some interval of time is given by the joint state entropy

$$\Gamma[Z(t), Z(t+\tau)] = \Gamma[Z(t)] + \Gamma[Z(t+\tau)|Z(t)]$$
$$= \Gamma[Z(t+\tau)] + \Gamma[Z(t)|Z(t+\tau)] \tag{4}$$

193

Biological organization

where

$$\Gamma[Z(t+\tau)|Z(t)]=-\sum_{\alpha,\beta}f[z_\alpha(t+\tau),\,z_\beta(t)]\,\log f[z_\alpha(t+\tau)|z_\beta(t)] \qquad (5)$$

is the conditional entropy, the uncertainty at time $t+\tau$ given the state at time t. The joint state entropy expresses the uncertainty as regards the system's state at a given time plus the uncertainty associated with the transition to some state at a future time or conversely. Accordingly $\Gamma[Z(t+\tau)|Z(t)]$ and $\Gamma[Z(t)|Z(t+\tau)]$ will be called the transition entropies. The joint state entropy is a function of τ.

The states of different compartments might not be independent. Thus

$$\Gamma[Z(t)]=\Gamma[Z_i^j(t),\,\bar{Z}_i^j(t)]=\Gamma[Z_i^j(t)]+\Gamma[\bar{Z}_i^j(t)|Z_i^j(t)] \qquad (6)$$

where $\Gamma[\bar{Z}_i^j(t)|Z_i^j(t)]$ is the conditional entropy of all compartments in the system other than c_i^j. This emphasizes that the state entropy of a compartment might be smaller when its relations to the rest of the system are taken into account. Naturally, the conditional term can be decomposed into further conditional terms.

Notice that $\Gamma[Z^*(t)|Z(t),\,Z(t+\tau)]$ represents selective and non-selective dissipation associated with loss of information about the environment. Accordingly this loss may be split into two terms

$$\Gamma[Z^*(t)|Z(t),\,Z(t+\tau)]=\Gamma_I[Z^*(t)|Z(t),\,Z(t+\tau)]$$
$$+\Gamma_{II}[Z^*(t)|Z(t),\,Z(t+\tau)] \qquad (7)$$

where the symbolism expresses the fact that the sum can be viewed as describing the concatenation of a noiseless machine with a noisy communication channel. $\Gamma_{II}[Z^*(t)|Z(t),\,Z(t+\tau)]$ is the equivocation. These do not include all the dissipative processes. Thus

$$\Gamma[Z(t),\,Z(t+\tau)|Z^*(t)]=\tfrac{1}{2}\{\Gamma[Z(t)|Z^*(t)]+\Gamma[Z(t+\tau)|Z^*(t)]$$
$$+\Gamma[Z(t)|Z(t+\tau),\,Z^*(t)]+\Gamma[Z(t+\tau)|Z(t),\,Z^*(t)]\} \quad . \qquad (8)$$

The first term on the right is the uncertainty as regards the correlation of system and environment. This will be called the correlation entropy. The second term represents the correlation between the environment at time $t+\tau$ and the environment at time t. This includes disturbances affecting the flow of information from the environment to the system, that is, noise associated with operations of the system. This is sometimes called the noise entropy. The third term represents selective and non-selective dissipation associated with loss of information about past states of the system. The fourth term represents mutability or inherent variability of the system.

The noise or correlation entropy may differ for different compartments. For example

$$\Gamma[Z(t+\tau)|Z^*(t)]=\Gamma[Z_i^j(t+\tau)|Z^*(t)]+\Gamma[\bar{Z}_i^j(t+\tau)|Z_i^j(t+\tau),\,Z^*(t)] \quad .$$
$$\qquad (9)$$

194

M. Conrad

In particular this is true if compartments respond on different time scales. This allows regulation.

The environment might give much less information about certain states of the system than conversely. In this case many system states are associated with a single environmental state. It may be possible to divide the state set Z into classes Z such that

$$\Gamma[Z^*(t)|Z(t), Z(t+\tau)] \approx \Gamma[Z^*(t)|Z(t), Z(t+\tau)] \qquad (10)$$

Thus

$$\Gamma[Z(t), Z(t+\tau)] = \Gamma[Z(t), Z(t+\tau)] + \Gamma[Z(t), Z(t+\tau)|Z(t), Z(t+\tau)] \qquad (11)$$

where $\Gamma[Z(t), Z(t+\tau)]$ will be called the reduced joint state entropy and $Z(t)$ represents the probability of each family of states of compartment c_j^i. The reduced joint state entropy is the uncertainty as to which of its families of trajectories the system is in, where further specification of the trajectory gives no more information about the environment.

Families of states correlated to the environment in different ways will be called functionally distinct. The members of these families will be called functionally equivalent. The latter must be distinguished from the (macroscopically) equivalent states. Later the notion of function will be given an independent definition.

The assumption of nonarbitrary families improves as $\tau \to 0$. This is because mutation may result in new associations between system states and environment states. The associations change as a function of τ because of the selective dissipation.

The joint state entropy is the uncertainty associated with the system's ensemble of trajectories. As $\tau \to 0$, this becomes the state entropy, or the uncertainty associated with the ensemble of states at some typical time. This is the ensemble of independent states. The number of possible states which the system can assume might be much greater than this. This is because the system may have certain definite trajectories or sequences of states. If these trajectories are out of phase the state entropy will be larger. This is reasonable since variability of the initial conditions should contribute to the state entropy. It is possible to define a state entropy (as opposed to a joint state entropy) over some natural time period— say a year. This will be called the path entropy. The ratio of the path entropy to the state entropy gives some idea of the complexity of the trajectories.

The state entropy is an information quantity, where information equals initial uncertainty minus final uncertainty. It is important that it must be manipulated as such, and not as a thermodynamic quantity. Flexible constraints do not play a role in ordinary thermodynamic systems. At equilibrium—when macroscopic

195

Biological organization

symmetry is as great as possible consistent with external constraints—most microscopic configurations are equivalent in terms of the properties characterizing the system, or are associated with small fluctuations. This is the reason why average values (constructed *a priori* with the help of an ignorance postulate) represent thermodynamic properties.

Flexible constraints are important in biological systems. Here the inequivalent states predominate. Many such states are compatible with a given energy condition—for example, consider the inequivalent states of a computer. In fact, it is this which forms the basis of the behavioral diversity of biological systems. The state entropy is determined from the frequencies of these inequivalent states, or states with macroscopically distinguishable effects.

The macroscopic diversity of a biological system is crucial apropos its behavior. In particular, it has a sharp effect on the overall energetics of the system. This is true even if macroscopic diversity is insignificant on a thermodynamic scale.

▶ *Efficiency*. The compartments of a biological system reproduce or are associated with reproducing units. According to the theory of evolution by variation and natural selection the environment classifies these units into two categories : those which actually reproduce and those which do not. Those which reproduce are fit.

The survival curve of a population, for example, species or community, depends on the reproductive success of the individuals in the population. The survival curve of a system will be described in terms of its energy as a function of time. According to the first and second laws of thermodynamics the slope of the survival curve is given by

$$dE/dt = dP/dt + TdS/dt - Td_iS/dt \qquad (12)$$

where E is the energy of the system, dP/dt is the influx of all forms of energy other than heat, T is the absolute temperature, S is entropy, and Td_iS/dt (≥ 0) is the dissipation [7]. This is true for open systems if $TdS/dt - Td_iS/dt = dQ^*/dt$, where this is the heat exchange with the environment per unit time, including specific enthalpies and heats of transfer of various components of matter.

Clearly d_iS is an inexact differential, or path dependent quantity. Therefore dP is also inexact.

The overall energetics of a system must conform to this relationship. However, the various terms in eq. (12) cannot be varied independently. For example, a decrease in the entropy of a system (dS/dt more negative) is a cost : all things being equal it lowers the slope of the survival curve. In general, however, all things are not equal and changing the entropy will usually affect the dissipation and the energy input. Similarly, an increase in the dissipation might lower the

196

slope of the survival curve, or might be coordinated to a more than compensating increase in the energy input. The dependencies among entropy, incoming energy, and dissipation depend on the detailed operations of the system and the character of the environment.

The efficiency of a system will be defined as its relative capacity to use the resources of the environment. Thus

$$u = \bar{E}/\bar{E}_{standard} \tag{13}$$

where u is the efficiency, $\bar{E}$ is the average height of the survival curve in a given environment and $\bar{E}_{standard}$ is the height of the survival curve of some standard system in the same environment. $\bar{E}_{standard}$ may be taken as the energy of some actual standard system or as the energy of some ideal biological system, that is, the maximum value of $\bar{E}$ in the given environment. The first possibility has some practical difficulty since the environment of different types of organisms is not arbitrary. The second possibility is difficult because it depends on an analysis of the inherent physical limitations of biological systems in the given environment. It is important that biological efficiency characterizes a system in a given environment only. This is reasonable since u should depend on the relationship between system and environment, that is, on the detailed operations of the system.

It is worth emphasizing the reason for this definition. Efficiency depends on energy input (chemical and radiant energy, work performed) and dissipation. These quantities do not characterize the state of the system or changes in this state—they are inexact. For example, some energy and entropy may be consistent with any compensating values of energy input and dissipation. Efficiency cannot be characterized in these terms.

It should also be clear that the maximum value of $\bar{E}$ cannot be taken as the energy accumulated by a free system performing the minimum amount of work and exhibiting the minimum amount of dissipation necessary for self-reproduction, without reference to the environment. In this case the relative capacity of actual biological systems to utilize the environment would always decrease with time. This is because the energy of the free system increases as long as energy flows into the system. This is not true for real systems. These must obey conservation laws (for example, chemical components, available space), resources inevitably become scarce, and therefore the growth of the system is limited. This inherent limitation is an essential aspect of evolution—otherwise there would be no natural selection.

The definition of efficiency, either in terms of a standard or ideal (biological) system gives an intuitive idea of its meaning. In practice, however, I think it must

Biological organization

be regarded as a primitive concept of biology. This is reasonable since it has the same properties as fitness. If efficiency increases (decreases) the height of the survival curve will increase (decrease). The converse is also true, unless the population is growing ($\bar{E}$ not defined) or the environment changes. If the population is growing an increase in efficiency results in an increase in the slope of the survival curve. In general, when a number of populations compete for the same energy input the one with the least unfavorable survival curve will predominate. Fitness is sometimes characterized by these relations to the survival curve.

The biological function of a trajectory will be defined as its relation to the survival curve. If the survival curve becomes more favorable the trajectory, or the system with this trajectory, is more efficient. I will assume that distinct families of trajectories (or states) are also functionally distinct in this sense, that is, have different relations to the survival curve.

▶ *Least state entropy.* In this section I will consider the relationship between the efficiency of a system and its state entropy.

The most efficient system is the one with the smallest possible reduced joint state entropy.

Contraction of the state entropy amounts to selection. This reduces the ensemble of trajectories or independent configurations. This may occur through a contraction in the size of the ensemble of systems in a population, or through a contraction in the ensemble of trajectories associated with each system. Not every contraction is appropriate in the sense that the systems selected may or may not be more efficient.

Least state entropy follows from the assumption that different families of trajectories are functionally distinct. This will be called the Gausse hypothesis, or competitive exclusion principle, since it is a generalization of the idea that distinct species cannot coexist in the same niche. In general, systems of different design are not equally well adapted to a given environment, that is, have varying degrees of efficiency.

The advantage of least entropy is a consequence of this variation in efficiency among the members of a population. Appropriate contraction in the reduced state entropy will remove the systems with the less efficient trajectories. Actually the system may assume appropriate trajectories in a given environment. The state entropy is not contracted as long as this potentiality exists. However, the potentiality is costly in thermodynamic terms. The system must construct and maintain machinery which it does not use. Systems which forego this machinery have an advantage.

198

In a definite environment—one with a definite sequence of states—the most efficient system will be one with only a single functionally distinct trajectory. However, such a system will be less efficient in an arbitrary environment. This is the interference between adaptedness and adaptability.

Even if the environment is definite the system will be subject to noise perturbation. Environmental noise may be absorbed by the equivalent configurations. This allows for loss of unnecessary information about the environment. The equivalent states do not contribute to the state entropy. The operations of the system are also subject to noise. The system must achieve some independence of these events without losing information about the environment. This is possible if the functionally equivalent trajectories are used to convert nonselective to selective loss of information. The functionally equivalent trajectories contribute to the joint state entropy. Reducing the joint state entropy by contracting the functionally equivalent states increases efficiency by reducing dissipation. However, this interferes with the selectivity of information loss.

I want to emphasize that least state entropy is applicable even if the reduced state entropy is not significant on a thermodynamic scale. The principle applies because of the functional character of the behavior of biological systems, regardless of the detailed nature of this character. It is not true because of the ordinary thermodynamic relation between entropy and work capacity of a system. In particular, function is dependent on constraints which control the use of energy, the coupling of processes, not on the thermodynamic entropy of the system *per se*. Reduction of the state entropy results in the most appropriately constrained systems. It may also result in the most inexpensively constructed systems.

Appropriate contraction in state entropy produces an increase in adaptedness relative to a given environment. Thus contraction in state entropy is a form of specialization of labor, that is, specialization for a less uncertain environment. Specialization of labor may also be associated with the evolution of novel adaptations. This means selection of a new set of systems, rather than just selection from a given set, or reduction in the variability of a given ensemble.

Thus efficiency can increase without a reduction in variability. This is possible when the system's design is selected or restricted for a given task. These restrictions are the constraints which control the operations of the system.

I will call such restrictions on design the internal specialization of a system. This must be distinguished from its external specialization, or the number of its dependencies on the environment. For example, external specialization increases

Biological organization

if the system requires a greater variety of building blocks, or can degrade fewer sources of building blocks. This allows the system to be more efficient since it can operate with a smaller amount of machinery. It also decreases efficiency since the uncertainty of the environment will be greater when the system depends on a larger number of independent factors. Changes in internal specialization at the top level of the ecosystem are determined by changes in the external specialization of its constituent species.

▶ *Internal specialization*. The internal specialization of a compartment should be related to its diversity and inhomogeneity in terms of subcompartments—to its degree of differentiation. This will be expressed in terms of the amount of selection required to describe the system given the types of subcompartments. The amount of selection, or information, equals some initial uncertainty minus some final uncertainty.

The final uncertainty is provided by the type entropy

$$\Gamma(c_i^j) = -\sum_r f(_j^j c_{\equiv r}^{j-1}) \log f(_j^j c_{\equiv r}^{j-1}) \tag{14}$$

where $\Gamma(c_i^j)$ is the type entropy of compartment c_i^j and $f(_j^j c_{\equiv r}^{j-1})$ is the frequency of subcompartments of type r. Types may fall into discrete classes or may be determined on the basis of overall similarity analysis. In the latter case refinement of the analysis must not change the distribution of types in each class.

Type entropy reaches a maximum, $\Gamma(c_i^j)_{max}$, when all frequencies are equal. The index of internal specialization, or degree of differentiation, is given by

$$\Gamma_d(c_i^j) = \Gamma(c_i^j)_{max} - \Gamma(c_i^j) \tag{15}$$

where the maximum type entropy provides the initial uncertainty. This is sometimes expressed as a ratio

$$D_i^j = \frac{\Gamma_d(c_i^j)}{\Gamma(c_i^j)_{max}} \tag{16}$$

giving a value between zero and one.

The index of internal specialization is a compromise statistic. It increases with an increase in the number of types, but only if the frequencies of these types are unequal. The index always represents the restriction on the frequencies given the number of types. It is important to distinguish this from redundancy, associated with restrictions on the relative frequency of components (symbols, for example) which are processed by the system.

The constraining relationships underlying redundancy differ from those underlying internal specialization. Redundancy relationships are characterized by the fact that the operation of only some parts is sufficient for the operation of the

200

system. This is possible if the relationships involve repetition of components, complication of the pattern of connection of components, multiple distribution of operations, or other forms of parallelism which allow for functionally equivalent states [see, for example, reference 3]. Redundancy changes relative frequencies of components in only an incidental way, and may in fact result in a decrease in the number of types.

Internal specialization relationships are characterized by the fact that the operation of many parts are necessary for the operation of the system, as in a series connection. Here the parts rely on one another, or are coadapted in the sense that they function in an environment of other parts. Such fitting relations become more intolerant as the internal specialization increases, since this amounts to an increase in the diversity and inhomogeneity of the internal environment.

Redundancy is associated with adaptability ; internal specialization with adaptedness. Actually the two categories of relationship are not independent since the redundant system must be associated with a coding apparatus.

Whether or not D_i^j is a satisfactory indication of internal specialization is a question that can be decided only on the basis of its usefulness. However, our intuitive picture of the complexity or internal specialization of biological systems is expressed by such indices. In the next three sections I will discuss the relation between the state entropy of a system and the uncertainty of the environment ; in the succeeding sections I will return to the relationship between state entropy and internal specialization. The particular indices discussed will play no essential role at present. However, the validity of the discussion can be assessed only through comparisons with biological nature, even if this is only in terms of an intuitive picture.

▶ *Static equilibrium.* In this section I will consider the relation between the variability of a biological system and the variability of its environment.

The reduced state entropy of a biological system tends to equal the state entropy of its environment plus the reduced correlation entropy. The former can be smaller if this is compensated by entropy associated with equivalent configurations.

The system utilizes information about certain environmental events and not others. For example, the system may require information about certain macroscopic conditions, conditions such as temperature, salinity, pH, in order to adjust to these. The system is also subject to other events, events associated with Brownian motion, radiation damage, and so forth. The system must cope with such noise, but this does not involve any transmission of information. In general,

Biological organization

biological systems contend with these various uncertainties by adjusting to them through adaptability mechanisms, blocking them with selective dissipation, or by tolerating a certain amount of damage. The systems which do this most efficiently, and which therefore have favorable survival curves, are the ones with the least state entropy.

Consider the form

$$\Gamma[Z(t), Z(t+\tau), Z^*(t)] \quad . \tag{17}$$

This may be expanded to give

$$\{\Gamma[Z(t), Z(t+\tau)] - \Gamma[Z(t), Z(t+\tau)|Z^*(t)]\}$$
$$+ \Gamma[Z^*(t)|Z(t), Z(t+\tau)] \equiv \Gamma[Z^*(t)] \quad . \tag{18}$$

The expression in the curly brackets represents the information which the state of the environment could provide about the state of the system at time t and $t+\tau$. The maximum value of this in any environment is defined as the capacity. Thus

$$\{\Gamma[Z(t), Z(t+\tau)] - \Gamma[Z(t), Z(t+\tau)|Z^*(t)]\}_{max}$$
$$+ \Gamma[Z^*(t)|Z(t), Z(t+\tau)] \geqslant \Gamma[Z^*(t)] \quad . \tag{19}$$

According to least entropy the capacity should be as small as possible. This follows since the capacity can always decrease through a decrease in the ensemble of functionally distinct states. The capacity takes on its smallest value when

$$\{\Gamma[Z(t), Z(t+\tau)] - \Gamma[Z(t), Z(t+\tau)|Z^*(t)]\}_{max}$$
$$+ \Gamma[Z^*(t)|Z(t), Z(t+\tau)] \to \Gamma[Z^*(t)] \tag{20}$$

where we require the equivocation to be small. The coding theorems of information theory suggest that this is possible and that the arrow can be replaced by an equality in the limit.

Capacity is the difference between the joint state entropy and the conditional joint state entropy. Thus a small capacity is consistent with a large joint state entropy. The actual value of the joint state entropy is important. According to least state entropy the system can be more efficient when this is smaller. The system will also be more efficient when the equivocation, the undesirable loss of information about the environment, is smaller. In fact a small value for the joint state entropy is not consistent with a small value for the equivocation. This can be seen by considering the basic character of coding theorems in information theory.

Equivocation arises from noise processes in eq. (8). According to the coding theorems this nonselective dissipation can be converted to selective loss of information about past states of the system. The system must be organized so

202

that spatial and temporal redundancies can be utilized for the selective recombination of states. Essentially the system must have families of trajectories— functionally equivalent but macroscopically distinct—such that noise perturbations do not cause the system to jump from one family to another. This convergence in state space requires encoding and decoding operations and therefore some increase in the selective dissipation associated with the joint state entropies. Thus the equivocation can be reduced only by making components (for example, enzymes) more reliable, in which case noise is by definition reduced, or by introducing a limited number of more reliable components and using these to convert nonselective into selective loss of information—essentially redesigning the system in such a way as to take advantage of these more reliable components. It is important that this redesign imposes certain costs: extra connections, associated with repetitions of operations in space, or delay, associated with repetitions of operations in time. The selectivity of the dissipation requires increased internal specialization at some level ; the increased dissipation is costly as well.

The dissipation will be smaller if the extra states required for error correction are as few as possible. However, there is a tradeoff between these extra states and the error correction. Thus there are very strong pressures on the actual values of the quantities in eq. (20). Changes in these pressures are important in the processes of biological organization and evolution.

It is important that the functionally distinct and functionally equivalent states are mixed in eq. (20) and cannot be separated. This is because functionally distinct states can arise from noise processes—for example, through mutation. These processes disappear as $\tau \to 0$. Then

$$\{\Gamma[Z(t)] - \Gamma[Z(t)|Z^*(t)]\}_{max} + \Gamma[Z^*(t)|Z(t)] \to \Gamma[Z^*(t)] \quad . \quad (21)$$

This can be written

$$\{\Gamma[Z(t)] + \Gamma[Z(t)|Z(t)] - \Gamma[Z(t)|Z^*(t)] - \Gamma[Z(t)|Z(t), Z^*(t)]\}_{max}$$
$$+ \Gamma[Z^*(t)|Z(t)] \to \Gamma[Z^*(t)] \quad . \quad (22)$$

In general $\Gamma[Z(t)|Z(t)] \geqslant \Gamma[Z(t)|Z(t), Z^*(t)]$. Assumption of the equality is reasonable since the distribution of functionally equivalent states in a family arise from noise processes and therefore depend on the character of the family, not on the macroscopic state of the environment. [This can be seen from the fact that

$$\Gamma[Z(t)|Z(t)] \geqslant \Gamma[Z(t)|Z(t), Z^*(t)] \Leftrightarrow \Gamma[Z^*(t)|Z(t)]$$
$$\geqslant \Gamma[Z^*(t)|Z(t), Z(t)] \quad .$$

The second inequality is possible only if $\Gamma[Z(t)|Z(t)]$ is large. However, according to eq. (10) $\Gamma[Z^*(t)|Z(t)] \approx \Gamma[Z^*(t)|Z(t)]$. Thus $\Gamma[Z(t)|Z(t)]$ can be

Biological organization

large only if there are families of states which give as much information about the environment as individual states, but which include states which are differently correlated to the environment. This is not possible. Notice that

$$\Gamma[Z(t), Z(t+\tau)|Z(t), Z(t+\tau)]$$

may become large as τ increases—in this case there is a certain arbitrariness in the assignment of states to families due to mutation. However, eq. (10) is not such a good assumption as τ increases.] Thus

$$\{\Gamma[Z(t)]-\Gamma[Z(t)|Z^*(t)]\}_{max}+\Gamma[Z^*(t)|Z(t)]\rightarrow\Gamma[Z^*(t)] \quad . \qquad (23)$$

Now least state entropy does imply that efficiency increases as the reduced state and correlation entropies become smaller. Accordingly

$$\Gamma[Z(t)]+\Gamma[Z^*(t)|Z(t)]\rightarrow\Gamma[Z^*(t)]+\Gamma[Z(t)|Z^*(t)] \qquad (24)$$

where it is understood that both $\Gamma[Z(t)]$ and $\Gamma[Z(t)|Z^*(t)]$ are determined independently.

The correlation entropy is a variable biological property. It must depend on the uncertainty of the environment given its past history. The correlation entropy is not an objective property of the environment ; rather, it depends on the system's capacity to recognize environmental states as signals and to extract these from noise. What is important are the system's mechanisms for making correlations as well as the relative time scales on which these mechanisms operate. Certainly these change in the course of evolution.

Correlations between system and environment are affected by all those processes which influence the flow of information from environment to system, that is, by processes contributing to the noise entropy. If the noise entropy is high the correlation entropy must also be high ; an increase in $\Gamma[Z^*(t)|Z(t-\tau)]$ implies an increase in $\Gamma[Z(t)|Z^*(t)]$, assuming that the system's capacity to recognize environmental states as signals does not change.

The noise entropy represents internal noise processes affecting the operations of the system. This can be split into various terms. Terms associated with compartments at lower levels represent reliability, or component failure. Terms associated with higher levels represent competition, that is, the fact that biological systems may have to cope with the uncertain behavior of other biological systems as well as with the external environment.

Mutation and competition decrease the correlation between system and environment. These processes certainly play an important role in evolution. However, the magnitude of the noise entropy is limited by the fact that information about the environment must be used for regulation. If the noise entropy is high the correlation entropy must also be high. If the correlation entropy increases the

204

regulative capabilities of the system must decrease, that is, the ensemble of independent families must increase.

Similarly, the state entropy cannot be indefinitely decreased at the expense of the dissipation term. This term represents processes which block out environmental states. For example, the processes of perception, of singling out the essential features of the environment, must involve some selective loss of information. As another example, the size of biological systems allows them to be independent of random fluctuations in the environment; here what are fluctuations in one system will not appear as fluctuations in any essential variables of another system [8]. Many protective mechanisms, such as rigid structures, play a similar role. However, certain information about environmental states is essential to the operations of the system, and the contribution of the dissipation term is limited by this fact.

The selective loss of information about the environment implies some variability in the microscopic states of the system, variability unnoticeable from the macroscopic point of view. The random details of the environment may induce variations of the system which essentially make no difference. Alternatively, loss of information about environmental states may be mediated by loss of information about past states of the system, implying that the equivalent states arise from equilibration (reset) processes. The dissipation decreases as the ensemble of equivalent states become smaller, but in this case the possibility for error increases.

It is also important that the system may interact selectively with certain features of the environment or it may be able to select the environment in which it operates (niche selection). This environment may change or may not correspond to the environment defined by the biologist. It may be convenient to include such specificity and avoidance reactions in the information loss term. In this case the information loss is not associated with an ensemble of equivalent states.

We can symbolize the above interpretations

$$\Gamma(system) \rightarrow \Gamma(environment)$$

where $\Gamma(system)$ is the state entropy and information loss of the system and $\Gamma(environment)$ is the state entropy of the environment plus the correlation entropy. The arrow represents a tendency, subject to the limitations discussed. This will be called a tendency for static equilibrium.

▶ *Levels of regulation.* Equation (20) can be written so as to take compartmental structure into account. Thus

$$\Gamma[Z_i^j(t), Z_i^j(t+\tau)] + \Gamma[Z_i^j(t), Z_i^j(t+\tau)|Z_i^j(t), Z_i^j(t+\tau)]$$
$$+ \Gamma[Z^*(t)|Z(t), Z(t+\tau)] \rightarrow \Gamma[Z^*(t)] + \Gamma[Z(t+\tau)|Z^*(t)]. \quad (25)$$

Biological organization

Similarly

$$\Gamma[Z_i^j(t)]+\Gamma[Z_i^j(t)|Z_i^j(t)]+\Gamma[Z^*(t)|Z(t)]$$
$$\rightarrow \Gamma[Z^*(t)]+\Gamma[Z(t)|Z^*(t)] \tag{26}$$

for the reduced state entropy.

Actually there are no preferred compartments in biology—to a greater or lesser degree states of interest may control complementary states. For example, it may be important to control the physical condition of the DNA molecule—to maintain it in a homeostatic environment. The physical properties of the molecule may be quite certain, but the sequence of nucleotides quite uncertain. This uncertainty is used to control states at the population level, and ultimately to control the physiological environment of the molecule.

Equation (26) can be written with all compartments on an equal footing

$$\sum_j \Gamma[Z^j(t)]+\Gamma[Z^*(t)|Z(t)]\rightarrow\Gamma[Z^*(t)]+\Gamma[Z(t)|Z^*(t)] \tag{27}$$

where

$$\Gamma[Z^j(t)]=\Gamma[\,\bigcap_i Z_i^j(t)|\bigcap_{\substack{i,k\\k<j}} Z_i^k(t)] \tag{28}$$

is the reduced state entropy characteristic of level j.

This can be further decomposed

$$\Gamma[\,\bigcap_i Z_i^j(t)]=\sum_i \Gamma_e[Z_i^j(t)] \tag{29}$$

where

$$\Gamma_e[Z_i^j(t)]=\frac{\Gamma[Z_i^j(t)]}{n}+\sum_{r=1}^{n}\left\{\sum_{i_1\neq i}\cdots\sum_{\substack{i_1\neq i\\i_r>i_{(r-1)}}}\frac{\Gamma[Z_i^j(t)|Z_{i_1}^j(t),\ldots,Z_{i_r}^j(t)]}{n\binom{n}{r}^{-1}}\right\} \tag{30}$$

is the effective state entropy of compartment c_i^j and there are n compartments at level j. This is just a linear combination of conditional terms, convenient for writing the equation. It assigns a unique number to each compartment.

The characteristic entropy of a level is the sum of the effective entropies of the compartments at the level given the states of all compartments at lower levels. This means that the characteristic entropy is the uncertainty with which sub-compartments of these compartments are connected given all information about their states in terms of fundamental components. The characteristic entropy can be taken per unit biomass.

Many conditional terms are unimportant. In particular this is true if the description of the system is restricted to level structure. Information certainly passes from level to level in a biological system. The composition of a level may be determined by the composition of other levels—for example, the genes provide

206

information about the composition of all levels. However, the possible states of different levels are often quite independent. Thus the sequence of nucleotides in the gene does not determine which of its possible states the cell, the organism, or the ecosystem is in. The state of a population, on the other hand, often depends on lower levels—on alternate modes of development or alternate modes of behavior. In the absence of such dependencies the different configurations of the population represent oscillations, unfavorable in terms of their survival curves, particularly if organisms are internally specialized and have long generation times.

In general a compartment or level can be regulated only to the extent that environmental uncertainty and noise are selectively destroyed or removed by the joint state entropy of other compartments or levels.

▶ *Network analogies; vector representation of biological organization.* Static equilibrium relations are tendencies which might be expected in the course of evolution, not any exact or rigorous condition. In the limiting case, however, they express a symmetry between system and environment which is reminiscent of the static equilibrium of forces around a point : equality of action and reaction, Kirchoff's laws, conservation of mass in hydraulic mass flow, and so forth. For example $\Gamma[Z^*(t)]$ can be analogized to an applied force, $\Gamma[Z_i^j(t)|Z_i^j(t)]$ to an elastic compliance, $\Gamma[Z_i^j(t)]$ to an inertial element, and $\Gamma[Z^*(t)|Z(t)]$ to a dissipative element. As another example, $\Gamma[Z^*(t)]$ is analogous to the mass flowing into a region in hydraulic mass flow. This must equal the sum of the mass leaving the region, the mass increase in the region, the mass loss due to leakage, minus sources of mass which are regarded as internal. These are analogous to $\Gamma[Z_i^j(t)|Z_i^j(t)]$, $\Gamma[Z_i^j(t)]$, $\Gamma[Z^*(t)|Z(t)]$, and $\Gamma[Z(t)|Z^*(t)]$, respectively.

Principles of static equilibrium always tie the elements of a network together. In the case of an information network, static equilibrium expresses the notion that impinging entropy will be filtered out, siphoned off by correction channels, or absorbed in the region of interest. If one region of the net becomes cool, in terms of variability, some other region of the net must become warmer.

It is important that the regions of high and low variability might be at different levels of organization. This can be represented in the form of a vector model, as illustrated in Figure 1. $\Gamma(environment)$ is represented on the abscissa and $\Gamma(system)$ on the ordinate. $\Gamma(system)$ can be described in terms of component vectors—for example, the $\Gamma[Z^j(t)]$—and the dissipation term, where the latter is assigned to the zero level. These are abbreviated as Γ^j, with $\Gamma^0=\Gamma[Z^*(t)|Z(t)]$. $\Gamma(environment)$ is also divided into (equal) components, each associated with

Biological organization

one of the system vectors. These represent the fraction of environmental uncertainty which each level might be expected to handle, all things being equal.
The individual vector sums may form any angle with the coordinate axes, but
according to static equilibrium, the total sum should always tend to a diagonal.
Vector sums pointing above the diagonal line are associated with controlling
levels ; those pointing below the line are associated with controlled levels.

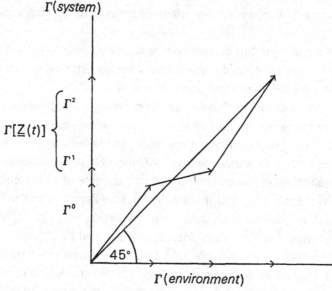

FIGURE 1
Vector diagram with two controlling levels and on controlled level

If types of compartments are taken into account the components of the environment vector can be weighted according to the fraction of biomass associated
with the type.

The main significance of the static equilibrium relationship is that it imposes a
certain total uncertainty on a biological system—the distribution of this uncertainty to the compartments must be consistent with this total. The covariance
of the vector model is at best an idealization, but the tendency of biological
systems to fulfil this condition should help to account for some of the tendencies
exhibited in the course of evolution. The quantitative correctness of the principle,
the extent to which the limiting values are reached, is not as important as
whether the total uncertainty can be treated as an approximately conserved
quantity. The validity of this hypothesis, as well as the validity of related assump-

tions about coding, reliability, information loss, absorbed variability, and so forth, can be determined only by constructing models of biological organization.

▶ *Allocation of the entropies.* The ecosystem is differentiated into subpopulations, each occupying some local environment. In what follows the discussion will be limited to a single species in a definite environment. We will suppose that the uncertainty of the environment is a given fact, and that the internal specializations of the species are also given. What is of interest are the factors which determine the allocation of adaptability to different levels of biological organization.

Efficiency depends on the intrinsic and extrinsic relations of a system, relations hard to analyze in general. However, if the restrictions on relations—the internal specializations—are fixed the efficiency can change only as a result of factors which interfere with these relations. This interference arises from adaptability, or the system's adaptedness to an ensemble of environments, or the environment in the long run. Thus I will assume that

$$U = U(\Gamma^1, \ldots, \Gamma^n ; D^1, \ldots, D^{n+1} ; \Gamma) + C \tag{31}$$

where U is the efficiency for given internal specializations, Γ^i represents the adaptability of level i, D^i is the internal specialization ratio characteristic of level i, Γ equals $\Gamma(environment) - \Gamma^0$, and C is a constant. Here we are ignoring the compartmental structure of biological organization and are characterizing levels as a whole. The notation would otherwise be somewhat more complex. U must be maximized subject to the constraint

$$\Gamma = \sum_i \Gamma^i \tag{32}$$

where $\Gamma^i \geqslant 0$ for all i.

Equation (31) does not include any detailed facts about the species or its local environment : the available energy, the number of required building blocks, the number of sources of building blocks, the thermodynamic costs of internal specializations, the topographic complexity of the environment, the complexity of the environment trajectory (ratio of path entropy to state entropy), and so forth. Such facts would have to be included if determination of internal specializations were part of the optimization problem. However, in the present case these facts are not necessary since we are assuming that the interference of adaptability with the function of the system can be described in terms of the internal specializations—it would be redundant to specify the facts which determine these specializations.

The C term is required in eq. (31) because the detailed relations underlying internal specializations are not specified. These are necessary in order to determine

209

Biological organization

U, the efficiency. C is constant if these underlying relations do not change. This is a reasonable assumption since the adaptability of a system, its statistical or control characteristics, might expand or contract within the framework of certain essential constraints. For example, changes in the variability of the gene pool might depend on minor physiological changes. Changes in behavioral plasticity might also take place relatively fast on an evolutionary time scale, without major changes in the differentiation of the organism or the nervous system. In contrast, the development of new types of adaptations, such as new reproductive or behavioral mechanisms, requires longer periods of time. Actually C might be roughly constant even in cases where the underlying details change somewhat. This would be true if the optimizing processes of evolution induce a certain parity among these details.

The D^{n+1} term represents the internal specialization of the ecosystem above the species level. The adaptabilities may interfere with the function of the ecosystem. This fact must be considered because of the assumption of a definite environment. If the function of the ecosystem is disturbed the state of the environment will not be re-established, the ecosystem will not be balanced, and the factors of selection will change. In general the local environment is not an arbitrary fact ; the local environment determines the internal specializations and conversely.

Equation (31) is the sum of two terms, the first changing with redistribution of the adaptabilities, and the second depending on an unknown process. If these processes are fixed, the maximization of U depends only on the distribution of the adaptabilities. These adaptabilities confer certain advantages on the system and also certain disadvantages.

Gain in efficiency results from the fact that the adaptability must approximate the uncertainty of the environment. This represents either functional states of the system (adaptability) or nonfunctional states forced on the system by the environment. These nonfunctional states cannot be avoided unless the uncertainty of the functional states exceeds the uncertainty of the environment. This must be so for those systems which survive. The advantage of adaptability is expressed in the constraint (32), except that the advantage deriving from any particular compartment may be associated with diminishing returns. For example, it may be difficult to couple the variability of some compartment to the control of certain processes ; or the compartment may not operate on an appropriate time scale ; or redundancies associated with the adaptability of a compartment may not be compatible with the overall operations of the system.

There are also certain losses associated with adaptability. This is the converse

210

of the least entropy principle. Usually efficiency will not increase with a contraction of the state entropy—the chances are that the most efficient systems
will be eliminated. Likewise, the chances are that the expansion of the state
entropy will decrease efficiency—most of the new variations will not be as fit as
the original system.

The possibility of variations which increase efficiency is important from the
point of view of evolution ; from the point of view of biological organization,
however, the rate at which efficiency falls off is important. This will be discussed
in the next section.

▶ *Complementary relations*. Adaptabilities of different compartments do not
interact, in the sense that the adaptability of one should not affect the adaptabilities of the others. Thus

$$U = \sum_i U_i(\Gamma^i ; D^i, \ldots, D^{n+1} ; \Gamma) + C \tag{33}$$

where U_i is the contribution associated with the adaptability of level i. Thus the
adaptability of a compartment has a complementary relationship to the internal
specialization of certain other compartments.

Each of the U_i results from a gain, associated with diminishing returns, and a
loss, associated with the fact that adaptability, involving variations of constraints,
interferes with the adaptedness of internal specializations. A system must be well
formed—its parts must be related in a way which is functionally integrated.
Rearrangements of parts are more likely to interfere with such relations as the
system becomes more differentiated, or as the relations among parts becomes
more restrictive. Any subsystem of the ecosystem also operates in a biological
environment which is more or less homogeneous or diverse. Variability is more
likely to reduce the adaptedness of a system in a differentiated environment since
the relationship between compartment and environment is usually more restrictive
in this case. For example, a cell may be highly differentiated internally, in which
case variability will be expensive. The cell may also function in an environment
(organism or society of cells) which is highly differentiated. It must be adapted
to the cells of this society, and this will also increase the cost of variability.

There are prominent exceptions to these generalizations. These are the special
organs of variability—for example, the genetic system, the immunity system, the
central nervous system. These organs have a more or less universal property since
they may be tailored or programmed for different modes of behavior. The essence
of this property is the isolation of certain crucial constraints from certain adjustable
constraints. In this case variability does not disturb the internal structure of the

211

organ as much as usual, although it may still disturb the relations between system and environment. However, the complementarity of internal specialization and adaptability holds in a broad sense since the system will be particularly sensitive to the conditions affecting the maintenance and construction of essential constraints. This increases the cost of variability at higher or lower levels of organization. For example, the operation of the genetic readout machinery requires a highly controlled cellular environment ; the operations of the central nervous system require a controlled environment as well as some highly differentiated cells.

Actually the cost of variability at the genetic level does depend on correlations among genes—the more coadapted the genes the less likely that an arbitrary variation will be viable. This is because such variability interferes with the function of compartments described by the genome. This includes the proteins of the cell, and through these the cell and the organism. The characteristics of organisms which produce coadaptation at the social and higher population levels are also reflected in the genome description in the sense that phenotypic correlations must be associated with genotypic correlations. Thus the cost associated with the variability of the genome depends in some way on the internal specializations of the compartments described. It does not depend on the internal specialization of the genome itself since the genome does not include its own description. Adaptability at the genetic level is also limited by internal specialization at the organism level in so far as this increases the reproduction time.

Variability at the top level of the ecosystem also has special relations to the ecosystem as a whole. The ecosystem must obey conservation laws. Thus variations in available energy must appear either as variations in the energy utilized by individuals or as fluctuations in the size of populations. The latter implies organism death. The cost of such fluctuations increases with the internal specialization of the organisms involved ; furthermore the growth rates of internally specialized organisms are slower and therefore the delay time before the environment can be fully utilized is increased.

Variation in the energy balance of individuals amounts to organism adaptability. If this is not possible, fluctuations can be avoided only by sacrificing energy to the detritus cycle. This means that the fluctuations occur in the detritus populations—here the organisms are simple, the growth rates high, and the cost of fluctuations relatively small. This sacrifice can be avoided to the extent that populations control the possible pathways of the flow of energy, taking advantage of any compensating variations in the environment. Control over possible pathways of energy flow implies correlations among organisms and therefore internal specialization at the top level of the ecosystem [9]. Indeed the stability of a foodweb often increases with an increase in internal specialization relative to the species composition of the ecosystem [10].

Gene pool variability favors the continuity of lineages. As such it stabilizes the ecosystem apropos its composition in terms of types. The internal specialization underlying foodweb stability reduces oscillations by reducing the competition term, that is, by reducing uncertainty of biological origin. As such it stabilizes the ecosystem apropos the frequency of types. Evidently there will be some interference between foodweb stability and the adaptability of the gene pool since genetic variability must be consistent with correlations at the top level of the ecosystem. These correlations are the result of community selection—they are a long run, not immediate advantage to the species involved. Accordingly, the genetic system may itself develop homeostatic features in order to preserve these long-run advantages. Biological systems often develop latent mechanisms of variability—dominance, suppression, inversion, and so forth—to evade the immediate consequences of this interference.

▶ *Models of biological organization.* In this section I would like to illustrate the preceding discussion with a definite model, based on the simplest assumptions.

Assume that the gain function exhibits diminishing returns and that this can be expressed as

$$G_i = 1 - e^{-\alpha_i \Gamma_i / \Gamma} \tag{34}$$

where $\alpha_i > 0$ for all i. Also, for simplicity, assume that the loss function is linear

$$L_i = \sum_j d_{ij} \Gamma_i \tag{35}$$

where we define

$$d_{ij} = f_{ij}(D^j) \quad . \tag{36}$$

The f_{ij} are (unknown) functions relating the (given) internal specialization ratio of level j to the adaptability of level i.

Evidently we must maximize

$$U = \sum_i (1 - e^{-\alpha_i \Gamma_i} - \sum_j d_{ij} \Gamma_i) + C \tag{37}$$

subject to the constraint

$$\sum_i \Gamma_i = 1, \, \Gamma_i \geqslant 0 \tag{38}$$

where Γ has been set equal to 1 for convenience. Note that

$$U_i' = \alpha_i e^{-\alpha_i \Gamma_i} - \sum_j d_{ij} \tag{39}$$

and

$$U_i'' = -\alpha_i^2 e^{-\alpha_i \Gamma_i} \leqslant 0 \quad . \tag{40}$$

213

Biological organization

Thus all the U_i are concave downwards. In this case efficiency is maximized if the system shifts from a curve U_j to a curve U_i only when the slopes of these two curves are the same. Accordingly,

$$U_i'(\tilde{\Gamma}_i) = \lambda \Leftrightarrow \tilde{\Gamma}_i > 0 \tag{41}$$

where λ is the slope and $\tilde{\Gamma}_i$ is the value of Γ_i when U is maximized. This shift will occur only if the slope of U_i at the point zero is greater than λ, and will certainly occur if it is. Thus

$$U_i'(0) > \lambda \Leftrightarrow \tilde{\Gamma}_i > 0 \ . \tag{42}$$

Combining eqs. (39) and (41)

$$\tilde{\Gamma}_i = -\frac{1}{\alpha_i} \ln \left(\frac{\lambda + \sum_j d_{ij}}{\alpha_i} \right) \tag{43}$$

where $\tilde{\Gamma}_i > 0$. Together with the constraint (38) this gives

$$-\sum_{\substack{i=j \\ \tilde{\Gamma}_i > 0}}^{n} \frac{1}{\alpha_i} \ln \left(\frac{\lambda + \sum_j d_{ij}}{\alpha_i} \right) = 1 \ . \tag{44}$$

According to eq. (42) the condition $\tilde{\Gamma}_i > 0$ implies that

$$U_i'(0) = \alpha_i - \sum_j d_{ij} > \lambda \tag{45}$$

and therefore $\tilde{\Gamma}_i = 0$ if the sum (44) is greater than one when λ is taken as $U_i'(0)$.

The values of λ and $\tilde{\Gamma}_i$ are determined by the values of the α_i and the d_{ij}. The latter can be expressed in a matrix

$$\begin{pmatrix} d_{11} & . & . & . & . & d_{1(n+1)} \\ . & & & & & . \\ . & & & & & . \\ . & & & & & . \\ d_{n1} & . & . & . & . & d_{n(n+1)} \end{pmatrix} \tag{46}$$

where the choice of entries amounts to particular assumptions about biological organization.

Note that increasing values of d_{ij} will decrease the maximum value of U, but that an increase in the number of levels will have the opposite effect. From the point of view of overall efficiency the cost of developing or amplifying another level may be offset by this fact. When level i is absent we shall set $d_{ii} = \infty$ and $d_{ij} = d_{ji} = 0$ for all j. In this case adaptability cannot be allocated to the level.

We shall make the following assumptions, in rough agreement with the previous section.

(1) Rearrangements of compartments are usually not affected by internal specialization of compartments at lower levels, and are affected most by the

214

internal specialization characteristic of their own level and the immediately higher level. Thus

$$d_{ij} \begin{cases} > 0, j=1 \text{ or } i+k \\ = 0 \text{ otherwise} \end{cases} \tag{47}$$

where k is the smallest integer for which $d_{(i+k)(i+k)} \neq \infty$, and we are ignoring the contribution of higher levels.

(2) It is useful to distinguish developmental or growth adaptability from other forms of physiological adaptability which allow the organism to adopt different modes of behavior. Levels with special organs of adaptability are an extreme example of this. Here eq. (47) is replaced by

$$d_{ij} \begin{cases} > 0, j=i+1, i+2 \\ = 0 \text{ otherwise} \end{cases} \tag{48}$$

and

$$d_{ji} > 0, j=i+1 \tag{49}$$

where i is the exceptional level. $d_{ii}=0$, since rearrangements do not affect the organ itself, but this is compensated by the fact that the adaptability of higher levels is restricted [eq. (49)]. Actually, differentiation of organ structure would also increase the restrictions on some compartments at the cellular level. The special organ, as a form of physiological variability, is subject to the same environmental influences as variability at the organism level [eq. (48), $j=i+2$] ; such variability is also restricted by the fact that it must be consistent with the internal operations of the organism [eq. (48), $j=i+1$], although this factor may not be as important.

(3) In the case of the genome (at level 1)

$$d_{ij} > 0, j > 1 \quad . \tag{50}$$

These probably become less important as j increases, but they can never be ignored because of the mapping between genome and other levels. $d_{11}=0$ since the genetic system is a special organ.

Consider the coefficient matrices for systems with two levels (gene and cell), three levels (gene, cell, and organism), and four levels (gene, cell, organ, and organism), called system I, II, and III, respectively. For the sake of a numerical example, take all nonzero coefficients equal to 0·5, and α_i equal to 2 for all levels. The matrices and corresponding vector diagrams are given in Figure 2. Notice that adaptability is mostly at the genetic level in system I, mostly at the organismic level in system II, and mostly at the organ level in system III.

215

Biological organization

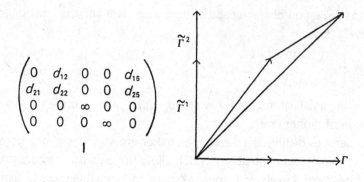

$$\begin{pmatrix} 0 & d_{12} & 0 & 0 & d_{15} \\ d_{21} & d_{22} & 0 & 0 & d_{25} \\ 0 & 0 & \infty & 0 & 0 \\ 0 & 0 & 0 & \infty & 0 \end{pmatrix}$$

I

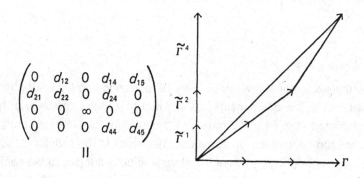

$$\begin{pmatrix} 0 & d_{12} & 0 & d_{14} & d_{15} \\ d_{21} & d_{22} & 0 & d_{24} & 0 \\ 0 & 0 & \infty & 0 & 0 \\ 0 & 0 & 0 & d_{44} & d_{45} \end{pmatrix}$$

II

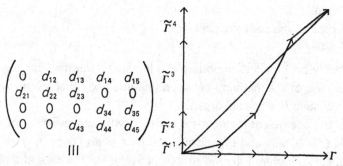

$$\begin{pmatrix} 0 & d_{12} & d_{13} & d_{14} & d_{15} \\ d_{21} & d_{22} & d_{23} & 0 & 0 \\ 0 & 0 & 0 & d_{34} & d_{35} \\ 0 & 0 & d_{43} & d_{44} & d_{45} \end{pmatrix}$$

III

FIGURE 2
Coefficient matrices and corresponding vector diagrams. The coefficient matrices are based on assumptions (47) to (50). System I is a unicellular organism, system II a multicellular organism and system III a multicellular organism with a developed organ level. $\tilde{\Gamma}^1$ is associated with the gene level, $\tilde{\Gamma}^2$ with the cell level, $\tilde{\Gamma}^3$ with the organ level, and $\tilde{\Gamma}^4$ with the (multicellular) organism level

216

The above discussion shows how models of biological organization, embodying definite assumptions, can be constructed within the present theoretical framework. Certainly the present assumptions are overly simple from the standpoint of biology. The constraint (38) is probably not so exact. I have not considered the possibility of changes in $\Gamma[Z^*(t)|Z(t)]$. Mutation, the coding aspect of the problem (the functionally equivalent states), and the time dependence of the entropies are ignored when $\tau \to 0$. The effects of these are lumped in α_i. This parameter may not be the same for all compartments, especially the organs of adaptability. The particular assumptions embodied in eqs. (47) to (50) are naive and the numerical values are arbitrary. The compartmental character of the problem has really been ignored. In particular I have not included adaptability at the population level—this would be important in the case of system I. Nevertheless, the model touches on some important processes underlying biological organization, and despite its artificiality, it is worth silhouetting it against the biological background.

▶ *Characteristic features of biological organization.* The biological world contains three main life plans, or kingdoms : the protista, the plants, and the animals.

The protista are characterized by three important levels of organization. These are the genetic, the cellular (organismic), and population levels. As an example, consider the bacteria. The bacterial cell exhibits various mechanisms of physiological adaptability—inducible enzymes, allosterism, multiplicity of components, and so forth. The bacterial population is also adaptable, as is evident from one of the standard methods of microbiology : the method of culturing bacteria. When the appropriate environment presents itself a particular bacterial type will be brought forth. This is possible because of the short generation times of bacteria. The adaptability may be based on the diversity of pre-existing types. These may persist in a spore form, capable of blocking or avoiding temporarily unfavorable environments. The adaptability may also derive from the genetic properties of bacteria ; from their haploidy, although there may be a certain amount of cytoplasmic lag associated with multiple nucleii ; or from their capacity to exchange genetic material through conjugation, transduction, and, perhaps, transformation in natural circumstances. This genetic interchange could occur in the context of hybrid complexes, allowing for the maintenance of types not adapted to the present environment. On the other hand, if an especially favorable adaptation is hit upon it could always be preserved by resorting to agamy. Inbreeding, agamy, diploidy, and periodic mutation (usually conferring an advantage on the larger, original population, and therefore lowering the

217

equilibrium proportion of mutants) damp unnecessary diversity, allowing for increased efficiency. These are the conservative mechanisms. In general, in a narrow environment, superfluous machinery, such as an unneeded inducible enzyme, puts a strain at a definite disadvantage, indicating that the requirements of least entropy may be very exacting.

The above picture, as regards the significance of sexual behavior, is not definitely established, although Ravin has presented strong arguments in its favor [11]. Even if sexual mechanisms are not so important at the present time, they may have played an important role in the original processes of bacterial speciation. The basic relationships, in the case of the other lower forms, are similar, although the details differ greatly. For example, protozoa range from inbreeders and asexual diploids to outbreeders and asexual haploids [12]. The first group, with specific adaptation, may have a short-term advantage, although this advantage can persist for a long time in a constant environment. It is interesting that asexual diploids possess a great deal of physiological adaptability due to the fact that multiple gene loci can respond differently to different circumstances. The second group, the outbreeders and asexual haploids, possess a great deal of evolutionary adaptability, but with a loss of high adaptedness. Haploid asexuals may, like bacteria, multiply rapidly and depend on mutation. Similar examples could be drawn from many other of the protista.

The higher plants exhibit a different pattern. Component vectors include the gene, the cell, and the population, and also the (developing) organism. Plant cell types fall into definite categories. The physiological adaptability of the various types is probably less than that found among the bacteria, at least while they are in the environment of the organism and after differentiation. However, as organisms, plants are remarkably adaptable. This is associated with their relatively simple morphology, which permits rearrangements, both in a developmental and comparative sense, to produce many different forms. This developmental plasticity, presumably controlled by quantitative alterations in a few growth hormones, allows the plant to accommodate itself to a variety of environments. Developmental plasticity is consistent with reproductive plasticity since new genetic information can be incorporated into the system without necessarily producing lethal results. Plants are capable of participating in extensive hybrid complexes, although there are mechanisms for preserving adaptations, such as ring chromosomes and agamy. Adaptations can also be maintained by extensive gene flow since this is the converse of the isolation conditions which favor speciation. In some plants the most apparent organs, flowers and fruits, are coupled to these

218

breeding systems. In certain cases facultative apomixis makes the rapid colonization of a new environment possible by genetically similar individuals, with a later burst of variability due to hybridization. However, the adaptability of plants at the population level is limited by their complexity and generation time. In particular this is true of the higher, more constrained forms. The patterns here sometimes parallel those in the animal kingdom. For example, consider the increasing protection for the embryo in both cases.

The component vectors of higher animals are the same as those of plants, except that the society (of organisms) is often interesting. The animal organism, like the plants discussed above, is a society of cells. This makes possible greater specialization at the cellular level, and therefore greater efficiency. However, there are more histological types among animals than among plants, although the number of types and their adaptability is definitely limited. The greater differentiation of animals is associated with the fact that they belong to higher trophic levels. From the thermodynamic point of view the environment becomes more demanding as trophic height increases. The differentiation of higher animals is associated with a closed growth system. This is not incompatible with rearrangements in a comparative sense, but it precludes the developmental plasticity characteristic of plants. For example, in mammalian species there are many possible arrangements of veins, fewer of arteries, and still fewer of nerves and muscles. However, the skeletal system is much more definite in pattern, and all in all the possibilities for ontogenetic rearrangements are limited. This lack of developmental plasticity in turn precludes the degree of genetic interchange possible in the plant and microbial worlds, for variety entering a complex and highly-differentiated structure is not likely to produce a viable result. In fact, this is attested to by the applicability of the biological species concept to, say, mammalian populations.

Thus, the gene, the cell, the developing organism, and the population cannot provide the required adaptability in higher animals. As a consequence special organs and organ systems must develop. For example, the nervous and neuromuscular systems make possible behavioral adaptability ; in turn, this extends the possibilities for motion, for social organization, and for specialization and homeostatic action at a new level. The other main system of adaptability is the reticuloendothelial system, the defense system of the body. The antibody associated with this system is representative of an ensemble of molecules, whether produced by an instructive or selective mechanism—just as the genetic molecule is representative of an ensemble of molecules. This is a clear instance of the entropy requirement of a homeostatic function. It might be remarked that systems with such high variability as the nervous or immune system are peculiarly subject to turning upon themselves, as in autoimmune disease. Indeed, adaptability is always associated with disease, if the latter is broadly conceived in terms of loss in adaptedness.

In sum, simpler forms can be variable at the genetic, cellular and population levels. Higher plants are constrained at the cellular and population levels, but compensate for this by variability at the (developing) organism and genetic levels. The higher animals are constrained at the genetic, cellular, organism, and population levels. Variability must appear elsewhere, in special organs and in social organization. In general increased specialization of one compartment of a biological system must be compensated by increased adaptability of some other compartment. As the internal specialization of subcompartments increases the cost of adaptability becomes prohibitive. A new level of organization must appear if it is advantageous for the increase to continue, or if it is advantageous for the system to exploit a more uncertain environment. The appearance of this new level may open the possibility for novel constraints, for internal specialization at the higher level, and therefore for the more efficient utilization of the environment.

Characteristic features of biological organization appear not for the sake of preserving any preferred entity of the biological world, but because they confer a preferred status in the matter cycle.

▶ *Summary*. Biological systems have a compartmental structure. The organization and variability of these compartments, or the degree of differentiation and uncertainty associated with them, are expressed in terms of certain entropy measures. Uncertainty is related to the adaptability of the compartment. The condition for most efficient operation of biological systems is discussed in these terms, as are various processes, such as computation, control, and competition. A principle of static equilibrium is proposed, according to which the uncertainty associated with the system approaches the uncertainty associated with its physical and biological environment. This is represented in terms of a vector model of biological organization. The relation only expresses conditions which biological systems tend to fulfil in the course of evolution. However, this makes it useful to treat the uncertainty associated with the biological system as a whole as an approximately conserved quantity. The distribution of variability to different levels of organization is discussed on this basis, and the model is compared to the background of biological fact.

M. Conrad

I would like to express my appreciation to Professor Behram Kursunoglu for the stimulating environment at the Center for Theoretical Studies. I would also like to thank Dr H. H. Pattee and Dr A. K. Christensen for helpful discussions, and Dr R. M. Williams for critical comments on the manuscript. I am indebted to Professor E. S. Castle for making it possible for me to begin a study of the subject.

This research was sponsored by the National Aeronautics and Space Administration under grant number NGL-10-007-010.

References

1. C. E. Shannon, *The Mathematical Theory of Communication* (University of Illinois Press : Urbana 1949).

2. W. R. Ashby, *An Introduction to Cybernetics* (Wiley : New York 1956).

3. S. Winograd and J. D. Cowan, *Reliable Computation in the Presence of Noise* (MIT Press : Cambridge, Mass, 1963).

4. M. Minsky, *Computation: Finite and Infinite Machines* (Prentice Hall : New York 1967).

5. K. B. Krohn and J. L. Rhodes, in (J. Fox, ed.) *Proceedings of the Symposium on Mathematical Theory of Automata*. p. 341 (Polytechnic Press : Brooklyn 1963).

6. H. H. Pattee, in (C. H. Waddington, ed.) *Towards a Theoretical Biology 2: Sketches* p. 268 (Edinburgh University Press 1969).

7. I. Prigonine, *Introduction to Thermodynamics of Irreversible Processes* (Wiley : New York 1961).

8. E. Schrodinger, *What Is Life?* (Cambridge University Press 1944).

9. M. Conrad, *Thesis* (Biophysics Program, Stanford University : 1969).

10. R. Margalef, *General Systems 3* (1958) 36.

11. A. W. Ravin, *Bact. Rev. 24* (1960) 201.

12. T. Sonneborn, in (E. Mayr. ed.) *The Species Problem: A Symposium Presented at the Atlanta Meeting of the American Association for the Advancement of Science* (The Association : Washington 1957).

The importance of molecular hierarchy in information processing

Michael Conrad
University of California

In this note I would like to confront some conventional concepts of information processing (selective dissipation) with the facts of molecular biophysics.

There are a number of theoretical computers which provide formalizations of the intuitive notion of effective (or algorithmic) information processing. The best known of these is the Turing machine. This is a finite automaton along with a memory space. The memory space consists of units whose states can be changed by the Turing machine—for example, it can write or erase symbols held by these units. The Turing machine can exert some control over which unit it accesses at a given time—for example, it may be able to move the units to the right or left. The way in which the Turing machine acts on its memory space is determined by the transition functions which characterize its finite automaton, for example, the tables which determine the next state and output given the present state and input.

The following facts about Turing machines are important.

1. There is a definite procedure for translating the transition tables of the Turing machine into networks of physical elements which give the desired behavior (for example, neural networks).

2. It is possible to program a universal Turing machine, that is, a Turing machine which can accept the transition tables of any special Turing machine as input (from its memory space) along with the original input data of this special machine, and transform this input data in the same way as the special machine. Any general purpose computer is universal in this sense.

3. The universal Turing machine can be translated into an actual machine according to the definite procedure in 1.

The definite procedure in 1 depends on two features of the components from which the actual machine is built. (a) The components are elementary units with definite transition functions or linkages of these. (b) The transition functions of linkages of elementary units can be derived from the individual transition function and the pattern of linkage, assuming that the inputs to each unit belong to its input set.

Each memory unit is also an elementary device—it receives input from the

222

automaton (or its reading head) and its state is a possible input to this automaton.

Now I would like to consider some notions about self-reproducing automata.
Von Neumann had the idea that universal constructors could be developed which
correspond to universal computers. Such a constructor may produce any machine
with whose description it is provided. If it is provided with a description of itself
it is self-reproducing—except that it does not construct another description.
This would lead to an infinite regress since in this case the description would
have to provide a description of itself. The system achieves complete self-
reproduction by xeroxing the description. Von Neumann exhibited a set of
elementary devices which in fact realizes this capability.

I should emphasize that in order to demonstrate the existence of such self-
reproducing systems it is necessary that the description serve as a program
(transition function) which prescribes the behavior of the system. This means
that the dynamics of the components which are being manipulated must be
suppressed—their behavior must be so constrained that they are completely
subject to a prescriptive computation process.

In sum, the existence of self-reproducing systems can be demonstrated. The
demonstration is based on the idea of a universal constructor which is provided
with a description. This description is a program, or a prescription for the behavior
of the constructor.

There is some resemblance between the overall process of self-reproduction
in such classical systems—involving transcription of and construction from a
description—and the biological processes of transcription and translation of
DNA. The resemblance, however, is superficial and does not confront the facts
of molecular biophysics.

The DNA is not a program or sequentially accessed control over the behavior
of the cell. This is because the biological process of translation does not correspond to the construction process. Translation just amounts to breaking the
energy degeneracy of DNA by coding it into the primary structure of protein.
This undergoes a spontaneous, energy dependent folding process. The function
of the enzyme (as regards catalysis or the control over energy transformations
and therefore the selectivity of dissipation in the cell) is determined by the three-
dimensional shape and charge distribution assumed in this folding process.
Naturally, there is some sequential accessing of different blocks of DNA during
different phases of the cell cycle. However, this sequential action is not comparable to the sequential action of the manipulable elements in a computer program.

223

Information processing

The manipulable elements in the DNA are accessed on a global basis, at least after they are translated to the primary structure of protein.

I shall call this mode of operation of the biological cell the hierarchical mode of information processing, as opposed to the single level mode characteristic of conventional computers and proposed constructors. Single level systems obey assumptions A and B—each unit in such a system reacts to defined outputs of other units, not to the global properties of collections of these units. In particular, they respond to (or access) distinct units in the memory space. This is because their interaction with different memory units is temporally distinct.

These features of single level systems are in sharp contrast to those of hierarchical systems. Hierarchical systems do not satisfy assumptions A and B since their mode of operation is based on the fact of hierarchy in molecular structure. It is not possible to program such systems in the conventional sense because the action of the program units in the description is not sequential and therefore cannot be combined according to a definite rule. Nor is it possible to manipulate the units in such a system directly in order to tailor its behavior or make it realize a given transition function. It is not a linked device whose transition function can be derived from the transition functions of its units, that is, it is not a device whose selectivity is increased simply by imposing certain initial conditions on given, unmodified subunits. It is really an elementary device with a complicated transition function.

These facts of molecular biophysics have implications for information processing in biological systems.

The forfeiture of prescriptive control means that a certain amount of information processing can be eliminated. This is possible to the extent that the information processing is not inherently sequential, that is, in so far as it can be realized by processes which really run parallel to one another. As a consequence, hierarchical self-reproducing systems may process information faster than general purpose single level systems which simulate them. This follows because the sequentially controlled single level device must execute a large number of temporally distinct (but perhaps very rapid) steps in order to reproduce itself. In contrast the hierarchical system acts on an energy basis. Essentially the process of self-reproduction in such a system is a constrained minimization of free energy The main constraint—at least the most manipulable one—is the DNA description. By and large this is not a sequential constraint and as a consequence many fewer (although perhaps not such rapid) steps are required for self-reproduction.

The hierarchical system may also be simpler than a special purpose single level

224

device which simulates it. (Of course no special purpose device has been proposed which is capable of producing more complex offspring.) Again this is associated with forfeiture of the capacity to design such systems to realize given transition functions—in this case by prescribing the appropriate pattern of linkages according to a definite procedure. Such prescription is always possible in the case of single level systems whose elementary units execute definite, usually simple functions. The rigidity of this function—the fact that selectivity of the system is modified by appropriate restriction of the possible inputs to the subunits and not by modification of the subunits themselves—often imposes significant topological complexity on the network which realizes the desired behavior. In contrast the hierarchical system is itself an elementary device with complicated behavior. The device is not built out of subunits with simple, definite functions ; rather manipulation of these subunits (at the primary level of structure) directly modifies the dissipative properties of the system. Again this is possible because the hierarchical system acts on an energy basis.

I want to underline an important point apropos this discussion of simulation. The use of a computer (single level system) to solve the equations of motion which describe a system is not the same as its use to simulate the system. Thus we might suppose that we can describe some system by certain state-to-state transition tables. Then the single level machine (tessellation automata, neural network, and so on) which realizes these transition tables simulates the system —at the cost, perhaps, of much more hardware. Alternatively, the single level machine may be designed to realize the input-ouput behavior of the original system by providing it with some transition table. This is an even looser notion of simulation, especially as the system may operate on an incommensurable time scale.

It should also be clear that the hierarchical mode of information processing is not the same as analog information processing. This is true even if the analog computer acts on an energy basis. In this case the choice and linking of components prescribes the physical law which the system will obey—the problem which the system must solve is communicated to it in this way. Such systems may communicate with digital devices or may be controlled by discrete information. However, they are never controlled by molecular folding and aggregation processes, that is, by the manipulation of a discrete substratum.

It is evident that the hierarchical system cannot be programmed like an ordinary computer—in order to plan its behavior we would have to consult physical laws and, of course, this plays no role in ordinary programming. Clearly such systems can only be 'programmed' on a trial and error basis—essentially through evolution by variation and natural selection.

In general such evolution processes are much more effective in hierarchical than in single-level systems. This is also a consequence of the absence of prescriptive control, for example, the fact that the DNA is a description of the primary structure of important molecules in the cell, not a prescription for the behavior of the cell. Ordinarily slight modifications in a computer program produce radically different behavior. This is because the modified program prescribes new behavior and there is no necessary relationship between this and the original behavior. This is not true in the case of the hierarchical system. This is based on energy primarily, and a modification of the description (constraint) will often produce only a slight perturbation to the behavior of the system. This increases the likelihood that single genetic changes will produce functional systems. This is unusual in the case of conventional computer programs.

The significance of this fact for the rate of evolution follows from fairly simple probabilistic considerations—essentially it is very much (astronomically) more probable for an evolutionary change to take place through a series of small steps each one of which is associated with objects (enzymes) which are to some extent functional or multifunctional than for it to take place in one jump. If the system as a whole is functional or has slightly increased fitness at each step the change may be amplified by population growth, thus increasing the likelihood of the appearance of the next step in the evolutionary sequence. In this case the evolution process is facilitated by transfer of function.

It is important that the effectiveness of this evolution process is itself an evolutionary property. This is because the extent to which a single genetic change perturbs the function of the enzyme depends on the extent to which it modifies its three-dimensional shape and charge distribution. If the enzyme is larger, that is, supports greater redundancy of weak bonding, the effect of a typical mutation may be attenuated. In this sense the topography of what has been called the adaptive landscape is itself an adaptive property.

Now I should like to consider some of the consequences of these molecular biological facts for concepts of information processing in general. According to the Turing–Church thesis any effectively computable function (function computable by some recipe) is computable by a Turing machine or by any of the machines which can simulate Turing machines. (All such machines compute the same class of functions—the partial recursive functions.) At least no algorithms have been found which cannot be expressed in this way.

226

M. Conrad

Certainly the Turing–Church thesis would not be so interesting if we could not communicate algorithms (or definite recipes) to actual machines, either through programs (in the case of universal machines) or by manipulating subunits directly. The question arises as to what happens when we deal with systems — systems such as the biological cell — to which we cannot communicate algorithms. Do such systems have properties which are essentially unattainable by systems with which we can communicate in this sense?

The fact of such systems is evident. All physical systems undergo some definite (but perhaps noisy) behavior. But this behavior is prescribed if and only if the transition function of the system can be communicated to it in a definite way. This is not true in most cases — most mechanical (physical) processes are not algorithmic processes. Nevertheless such processes may involve selective control over transformations of energy-selective dissipation. This is true, for example, apropos the elementary units of an ordinary computer (resets, 'or' devices, and so on). The information processing in such devices does not proceed according to a recipe that can be communicated to them. Rather, recipes prescribe the way these devices are linked together.

In the case of the biological cell these elementary processes are much more complicated ; furthermore, novel processes are often built by modifying the elementary processes directly rather than by combining them in different ways. Essentially the most important information processing in biological systems does not proceed according to definite recipes. Some rule (Turing program, for example) may generate the behavior of a biological system, but this does not mean that the biological system follows (embodies) this or any other rule (as opposed to obeying a physical law).

According to the Turing hypothesis, at least in so far as it has been interpreted as a link to psychological (or more broadly biological) processes, this distinction is of no fundamental significance — it is always possible to recapture the behavior of a system which processes information in the hierarchical mode by a system which processes information in the single level mode. However, this sense of recapturing or simulation is completely formal — it ignores entirely the spatial and temporal aspects of information processing.

I think, in view of the preceding considerations, that it is reasonable to conjecture that it is impossible to simulate a hierarchical system by a machine to which we can communicate algorithms (single level machine) without distorting its rate of operation or the amount of hardware which it requires. This tradeoff arises from the fact that it is necessary to pay for the constraints which make it

227

possible to prescribe a system's behavior. It is important that the sacrifice of this feature is compensated by the adiabatic modifiability of function which facilitates evolution processes.

This conjecture is evidently significant apropos the molecular processes underlying cellular self-reproduction and evolution. Does it have ramifications for higher-level biological processes, processes possibly based on similar underlying mechanisms ? I think that our approach to such processes ought not to be overly restricted by concepts of information processing which derive from the single level devices with which we are familiar.

I would like to thank H.H.Pattee and M.A.Arbib for stimulation in this general area. This work was supported by N I H Fellowship Number 1 F02 G M43960-01 from the National Institute of General Medical Sciences.

What can we know about a metazoan's entire control system?: on Elsasser's, and other epistemological problems in cell science

S. Kauffman
University of Chicago

Serious students of cellular and developmental biology confront what may well be the gravest epistemological problems ever faced by scientists. These are direct consequences of the immense, ordered biochemical complexity of organisms.

One of the most extensive discussions of the epistemological problems confronting cell biologists has been supplied by Elsasser whose views have had a rather wide audience [1, 2]. Elsasser is greatly to be admired for his insistence that biologists confront the epistemological problems of their science, and for his efforts to analyze those problems, which he feels center in our incapacity to know the quantum microstate of an organism. While I am sympathetic with his effort, I feel he misconstrues the epistemological consequences of our failure to know completely the organism's quantum microstates. More importantly, he and others have rather ignored the epistemological problems posed by the very advances of molecular biology, which center on how and what we can explain of the global behaviors of entire cellular control systems when we cannot know in detail all the components and interactions. Elucidation of this problem must surely be an integral part of increasing the capacity of molecular biology to cope effectively with the developmental biology of metazoans. I shall begin by analyzing Elsasser's argument, then turn to the problems facing extension of molecular biology to metazoan development.

The thesis Elsasser wishes to establish is the semi-autonomy of biology, by which he specifically means that organisms exhibit regularities which cannot, in principle, be deduced from quantum mechanics. He argues, in brief, for a radical irreducibility of biology to physics. The major problems Elsasser confronts in trying to maintain this claim are that he wishes also to assert that physics, specifically quantum mechanics, applies in entirety to organisms, and that no behavior of organisms contradicts physical law. In trying to establish that this autonomy of biology is possible, Elsasser uses two central concepts : Bohr's Generalized Complementarity and his own Principle of Finite Classes.

Bohr's Generalized Complementarity states that we cannot measure precisely all details of the microvariables, for example, electron spins in any single organism, without killing it due to the energetic aspects of the measuring process. This restriction restricts the precision of prediction of the organism's future behavior.

Elsasser then stresses that quantum mechanical predictions are never of individual events, but of probabilities of events, and make operational sense only if construed as predictions about frequencies of events within a given class of identical members, for example, atoms [1, p. 13]. He speaks of a set of atoms or molecules of the same composition and same quantum state as a fully homogeneous class, and says that measurement on homogeneous classes is 'an integral part of the measuring process in atomic and molecular physics. . . . The fully homogeneous class corresponds to a state of maximum definition or detail (and is) the tool of maximum predictability.' [1, p. 25]. Such homogeneous classes define a 'domain of verifiability' of the theory. Such classes are to be obtained by selection of appropriate subclasses from an initial inhomogeneous class. And, following von Neumann, since 'nothing prevents us from assuming the classes of atomic and molecular physics are of infinite membership . . . we can always procure homogeneous class of infinite membership. (Consequently) . . . it is no longer necessary to consider inhomogeneous classes by themselves. Once all the homogeneous classes into which they can be analyzed have been prepared, one can go back by suitable mixing. The properties of the inhomogeneous class can be readily calculated once those of the homogeneous class are known' [1, p. 29].

Next Elsasser claims that, in radically inhomogeneous systems such as a species of organisms in which the combinations of possible quantum states are enormous, the procedure for procuring a homogeneous class by selection of a finite number of successive subclasses may fail : 'we may quite simply run out of specimens during the process of selection before we have reached a point (of) adequate homogenization' [1, p. 36]. Such a class is called a class of finite membership, or a finite class. He next observes all organisms form finite classes, which leads him to 'non-Neumannian axiomatics in quantum mechanics, namely one in which one postulates that all classes in biology are finite'. This is Elsasser's Principle of Finite Classes, an irreducible inhomogeneity as a characteristic property of classes in biology.

In addition to difficulty in predicting due to our inability to describe in detail a single organism or compensate for that ignorance by obtaining a very large homogeneous class, Elsasser claims that cascading of errors through multiple feedback loops in metabolism will magnify and delocalize those errors, rendering prediction even more difficult.

230

S. Kauffman

Elsasser wishes to use the principle of finite classes to claim (1) that such inhomogeneous classes exhibit (organismic) regularities which have no equivalent in related homogeneous classes [1, p. 48] ; and (2) that these regularities are not deducible from quantum mechanics [1, p. 52]. The latter would render biology 'semi-autonomous' from physics. He thinks his non-Neumannian axiomatics will allow this second claim. 'J. von Neumann gave a formal proof . . . that quantum mechanics . . . is incompatible with the existence of regularities other than those which are deductively derivable from quantum theory itself.' Von Neumann's argument proceeded by showing that in a homogeneous class of infinite membership, quantum mechanics determines *all* class averages uniquely and permits precise predictions of these averages for future times. 'Thus any regularity which gives rise to predictable class averages must be . . . identical with those given by quantum mechanics.' On these grounds, 'biology would be purely mechanistic. . . . The solution lies on the change of an axiom ; the transition from infinite homogeneous classes to a finite universe' [1, p. 49]. 'Semi-autonomous biology makes sense only if it says : there are observable regularities which are intrinsically impossible to deduce in their entirety logico-mathematically from the laws of physics. . . . We may break up the conditions for lack of adequate prediction into two parts :'

1. Since we propose to operate within a finite universe of discourse, we may assume that if we hypothetically assign to any system event occurring in reality one specific microstate, then the number of microstates of any class in this universe will be negligible compared to the number of microstates which are theoretically possible : any actual microstate is as a rule an immensely rare event among all possible ones. This condition implies that the formation of all possible class averages (inclusive of averages over all microvariables, since a complete set of averages is a prerequisite for von Neumann's proof) becomes devoid of operational meaning for some of these averages.

2. The variability of microstates must not average out as they do in homogeneous systems and classes. Instead, this variability must modify a cascade of feedback cycles, ultimately influencing the macrovariables. If the macrovariables show regularities, as of course they do, then we may be able to use these to infer the properties of the particular microstates which are required to generate these macrovariables. If, however, the members of the class are radically inhomogeneous at all levels of their organization, this 'retrodiction' may be unsuccessful in the sense that a large number of microstates, each having characteristics rather different from any other, can be equally compatible with

the given macrovariables. Since on the other hand the experimental assessment of a microstate so far as it can be achieved is radically destructive, this very limited retroduction does not help us sufficiently to allow prediction of macrovariables for other members of the class [1, p. 75].
Elsasser goes on to say that some class averages can be found, and these correspond to the deterministic, 'engineering' mechanistic aspects of biology. However, since not all class averages can be found an irreducible and inseparable mixture of inhomogeneity remains. Nevertheless, these 'non-mechanistic' features can exhibit regularities, which are not deducible from quantum mechanics because the relevant class averages cannot be found. Thus biology is at least semi-autonomous. The extent to which an organism's regularity is due to stability properties of the deterministic aspects, and how much is due to regularities of the inhomogeneous semi-autonomous admixture, is, he says, unclear [1, p. 107].

There is, I think, something odd in the conclusions reached. One has a lack of feeling for just what kind of organismic order is due to the admixture of inhomogeneous finite classes. The unease is not aided by either Elsasser's failure to provide a single clear example, by assertion that this poses an intuitive paradox which hearkens scientific advance [1, p. 114], or by his admission that an unclear portion of the regularity of organisms may be due to 'deterministic' engineering aspects depending upon either high energy stable structures like DNA, or to those class averages which can be formed [1, p. 107].

Assuming for the moment that he is correct that there may be regularities not deducible from quantum mechanics, the only real help he gives towards their character stems from his second condition for autonomy of biology 'is that the variability of microstates must not average out, but instead due to cascades of feedback loops, must ultimately modify macrovariables' [1]. Here he argues that each macrovariable is consistent with very different microvariables such that limited retroduction from one macrovariable gains little of predictive value for other macrovariables. But consider his argument. If one macrostate corresponds to very different microstates, yet microstates form a space of states, each with similar neighbors, this implies that a microstate corresponding to one macrostate has neighbors corresponding to others. But since flow and uncertainty among microstates is greatest for those separated by very small energy steps (as he later argues) then 'noise' will cause passage among the neighboring microstates associated with different macrostates. Thus, if his argument for the conditions for regularities not deducible from quantum mechanics succeeds, it succeeds in

232

showing the *lack* of order among macrovariables in time, and helps not at all in understanding any regularity due to radical inhomogeneity. On the other hand, if a simply connected region of neighboring microstates corresponds to a given macrostate, which characterizes some regularity of an organism, he fails to show that we cannot explain the restriction to this class from quantum mechanics.

But the most important criticism I wish to make is that his argument is a *non sequitur* of a glaring sort. He wants to use his principles of finite classes to show that there can be regularities which are non-deducible from quantum mechanics, but that principle achieves nothing of the sort. The epistemological importance of an infinite homogeneous class is that statements of quantum mechanics are made about frequencies of events relative to a class. As he notes himself, such classes are the tools of *verification* or *falsification* of the predictions of quantum mechanics. If one is, contingently, restricted to a small finite sample because the existing number of specimens is less than the number of possible quantum states or microstates, *all* that logically follows is that we cannot verify or falsify all the corresponding predictions deduced from quantum theory. It in no way follows that any regularity which *might* be found in the inhomogeneous class is *not* in principle *deducible* from quantum mechanics. Formally, the principle of finite classes allows us to assert :

(1) Not all statements deducible from quantum mechanics describing regularities are verifiable or falsifiable. From which we may legitimately conclude :

(2) Some statements deducible from quantum mechanics describing regularities are not verifiable or falsifiable, that is, we are unable to decide if the regularities obtain or not. (2) is consistent with the case in which *none* of the regularities occurs, but we cannot verify this failure. Therefore (1) or (2) do *not entail* :

(3) There *are* some regularities the description (deduced from quantum mechanics) of which is not verifiable or falsifiable.

It does follow from (2) that there *may* be some regularities the description of which is not *verifiable* or *falsifiable*. But it does *not* follow that those descriptions are not *deducible* from quantum mechanics. The fallacy of Elsasser's argument can be brought home by considering that his characterization of a class as finite or not depends upon the empirically contingent relation between the number of existing examples of a class being less than the number of the potential microstates of a single member. Suppose, then, there were exactly three oxygen molecules in the universe, a possibility not excluded by quantum theory itself. These three would form a finite class on Elsasser's definition and his arguments must carry through here as well as with organisms. But surely we could utilize the wave equation for O_2 and deduce from that theory, as Elsasser himself says, predictions about thousands of spectral lines of O_2. Failure to verify statements about intensities of lines might follow from restriction to a finite class. But surely it would not follow that any regularity which happened to show up in the three O_2 molecules, for example, where the spectral lines are, would *not* be predictable by deduction from the wave function for O_2.

Elsasser seems to have gotten confused by the following line of thought. He reasonably suggests that we may use study of members of a class to make predictions of other members of that class [1, p. 75]. In particular, if we have a large enough sample of the class, we can form reliable class averages. We may then 'deduce' or predict in a trivial sense of deduce or predict that other members of the class would be characterized by the same averages. If it were the case that deduction of predictions in quantum mechanics necessarily required empirical study of reliable class averages of an ensemble of molecules or system, in order to be made, then in cases where class averages cannot be formed, deduction would be impossible. However, he is wrong that forming class averages is an integral part of deductions of regularities from quantum mechanics, as his own example of deduction from the wave equation for O_2 shows ; rather class averages are necessarily required for verification of deductions. Therefore, he does not show such non-deducible regularities might occur. Perhaps the form of his argument which comes closest to carrying through is his claim [1, p. 52]. 'At the same time, on making the available samples into representative samples of exceedingly rare events, one denies the possibility that all observable regularities can be deduced as logico-mathematical consequences from physical theory. For, in order to give this deduction an operational meaning, it would again be necessary to have a vastly larger set of samples than are ever available in actuality. Hence, in sufficiently inhomogeneous classes, the existence of such non-deducible regularities need not lead to a logical-mathematical contradiction with physics.'

Systematically, his argument is :
(1) Not all regularities can be deduced from quantum mechanics. Therefore,
(2) Some regularities cannot be deduced from quantum mechanics. Unfortunately, while the syllogism is valid, the antecedent is false. Failure to find a large homogeneous class only implies :
(3) All regularities deducible from quantum mechanics are not verifiable or falsifiable.

But (3) is not (1). Given (3), he attempts to support (1) by arguing that to give the *deduction* operational meaning, a large homogeneous class would be required, but cannot be had ; therefore the deduction is without operational meaning. That is, somehow, the deduction cannot be made.

Now, the operational theory of meaning was not, and is not, applied to *deductions*, and deductions do not have operational meaning. Presumably it is the deduced statement which Elsasser insists must have operational meaning. But it is just the possibility of *deducing* statements describing regularities of behavior from quantum mechanics which is in question, and surely such deductions can be made from, for example, a wave equation. If so, his antecedent, (1), is false and the syllogism fails.

At any rate, the operational theory of meaning is a very slim reed to support a whole new semi-autonomous biology with mysterious organismic regularities due to inhomogeneous classes. It is an old philosophic saw that a verifiability or falsifiability theory of meaning is severely called into question by asking how to verify the statement : 'Only statements which are verifiable are meaningful'. A thoroughgoing operational analysis of meaning has fared equally poorly ; for example, modify the above phrase to read : 'Only statements for which there are measuring operations tending to verify them, have meaning'. Criticisms of strict operationalism, as well as a straightforward positivistic verification-falsification theory of meaning can be multiplied *ad nauseam*.

Even assuming, for the sake of argument, the adequacy of operationalism, Elsasser's argument fails. The example of the finite class of two O_2 molecules allows us to examine a second interpretation of Elsasser's argument. Over any brief period of time the three molecules will occupy only a small region of their potential phase space. For that duration, the three may be characterized by certain regularities due to averages over the restricted region of phase space they happen to occupy. What is the status of such regularities ? It can surely be deduced from the wave equation *if* the molecules happen to be located in a given region of phase space ; those particular regularities will characterize them. What cannot be deduced *a priori* is the happenstance of their temporary restriction to this region of phase space which depends on the system's initial quantum state and rate of movement in phase space. Now, if the description of these regularities is *itself* operationally meaningful, then we know that the system is in one of a set of regions of phase space each of which, we can deduce from the wave equation, is characterized by these particular regularities. Any such regularities are surely deducible from quantum mechanics. Alternatively, there may be no subsets of phase space which would have the observed regularities as averages, in which case, those regularities would contradict quantum mechanics. Then, if the regularity is describable by an 'operationally meaningful' statement, that statement either accords with or contradicts quantum mechanics. Only if the description is not 'operationally meaningful', is the regularity, in Elsasser's odd sense of deduce, *not* deducible from quantum mechanics. But then we cannot *observe* the regularity, for to do so would render its description operationally meaningful ; and 'deducible' from quantum mechanics. Thus Elsasser fails to establish that observable regularities not deducible from quantum mechanics can exist.

Based on these arguments, I confess I think Elsasser has seriously misconstrued the central epistemological problems in cell biology and the source of order in biological systems in seeking it in semi-autonomous organismic regularities. It seems far more reasonable that the more common view is closer to the truth ; namely, order comes from the complex interworkings of the more nearly deterministic, mechanistic aspects of organisms. These would be due either to energetically stable structures like D N A, or to reasonably deterministic class averages, such as an enzyme's activity as a function of the fraction of precursor and product pools in acceptable microstates. Understanding such behaviors need not require keeping track of all microstates. Where such stability through averaging or energy fails, noisy fluctuation among energetically similar microstates should occur, and we reasonably expect that such tendencies will be for disorder, not order. Control processes seem likely to have evolved to use the more deterministic aspects of components' behaviors. It seems not entirely implausible that many or most of these 'mechanistic' aspects of individual components interactions may be sufficiently stable to be discovered and also to have physical explanations for them elaborated without knowing all possible microstates. If we are fortunate enough to learn how organisms interconnect these 'deterministic' behaviors of components to achieve stability and ignore the details of thermal and other noise in their orderly behavior, perhaps we can ignore its details too, thereby reducing our incapacity to explain how organisms work when Bohr's generalized complementarity holds. Indeed the most misleading aspects of Elsasser's statement of our epistemological problem is the suggestion that we either need to, or ideally would seek to know in detail the quantum-microstate of an organism in order to predict all its future behaviors. But our purpose surely is not to be able to deduce all possible true statements about regularities in organisms ; it is to explain how organisms work.

This brings us to consider the central epistemological problems associated with trying to understand how the more deterministic processes in cells interact to explain their behavior. After all, it is just such 'engineering'-like features of an organism's metabolic system which molecular biology is so successfully exploring, and whether we wish it or not, we must soon confront the epistemological problems its very success engenders.

An organism, fundamentally, is an autocatalytic, open thermodynamic system operating far from equilibrium and often far from even steady states. Specific catalysis and the consequent possibility for poisoning of catalysis, coupled with displacement from equilibrium, constitutes the physical-chemical basis of controlled onset and cessation of metabolic flows.

E. coli may code for 2000 proteins and deal with a thousand or so small metabolites. Mammalian cells may code for as many as 10^5 or 10^6 distinct genes. Of these, many are structural genes, but an untold number are surely control genes of various kinds whose major functions are to control one another's activities, as well as the activities of structural genes.

A consequence of the possibility of allosteric control is that an input molecule, for example an inducer, need bear no structural relation to either the product of the gene or the enzyme whose activity is modified. Therefore, when we consider the possible topologies of control between control genes, and from control genes to structural genes, this important constraint is missing. The only constraints on control structures we know are these two : (1) a tendency for enzymes specific for one metabolic pathway to be jointly controlled ; (2) a tendency to form negative feedback loops such as end product repression and inhibition—structures of obvious survival value. Above this structural level we have virtually no idea what the topology of control connections among genes and their products look like. How metabolic control systems should be built to obtain an organism's desired cell types, maintain them stably, achieve reliably the processes of differentiation among those cell types, and achieve reliably the correct spatial orderings of cells and structures, are questions to which we have only the vaguest clues. Almost the only claims which seem reasonably secure, if weak, are that random mutation will be expected to randomize the cross-couplings in the control topology ; and that selection will have chosen those with sufficient order for its purposes.

The heritage of anatomy, through biochemistry and current molecular biology, enjoins us to seek to explain the detailed order of an organism's metabolic workings by discovering the detailed character of components and the detailed control connections between them. However, while it seems reasonable that current approaches may allow us to establish the kinds of molecules which constitute the components of a cell, and also provide physical-chemical explanations of how each type of molecule interacts with its topological neighbors in the control structure, it is far less obvious that these approaches will allow us to trace all the control connections among a possible 10^6 components. Indeed, an injunction to analyze in detail the control connections among 10^6 components is mere cant. While logically possible, and, I believe, not ruled out by epistemological considerations such as Elsasser's (as discussed above), such a task would clearly pose Herculean difficulties—difficulties sufficient to warrant the rubric of epistemological crisis.

It is therefore germane to ask why we suspect we might need to trace all those control connections. An analogy of an organism to a special computer helps to state the issue. A general computer becomes equivalent to a special computer when given a specific program. Very small alterations in a program, for example, reversing the order of two cards, commonly produce enormous differences in the behavior of the general computer—that is, produce different special computers which behave very differently. To understand the detailed behavior of a particular program requires detailed elucidation of its flow diagram. If we could see only the wiring diagram of an equivalent special computer, we feel we would need to follow through its entire anatomy and rules to understand in detail the program the machine realized, for we know small changes in that wiring diagram could wield enormous changes in the behavior of the machine. Similarly, in organisms, we expect that small changes in the topology of metabolic control connections could easily wreak enormous changes in the organism.

Stated briefly, the problem pointed up by the computer analogy is this : for highly-ordered systems such as cells and organisms, some aspects of their behaviors are very sensitive to even small changes in the topologies of control connections, other behavioral properties presumably are relatively structure-insensitive. We currently have almost no idea what the appropriate properties are, or what properties belong to which class. Furthermore, we have almost no idea which behavioral properties of complex control systems evolution has acted on, selecting survivors. However, overwhelmed by the staggering, detailed way specific molecules and subsystems work together in metabolism, we seem to feel a strong prejudice that the behavioral properties of the control system which selection has worked upon and chosen are themselves as improbable as the detailed molecular harmony we see. That is, we seem disposed to think that selection has operated on behaviors of complex control systems which are very structurally sensitive and therefore highly improbable. This in turn implies that to

238

understand behavior of such structurally sensitive systems will require detailed analysis of a great portion of its control structure, just the injunction which confronts us with grave epistemological difficulties.

Yet what kinds of information can we reasonably expect to obtain about the topology of cellular metabolic control systems ? Eventually, perhaps, the entire story, but until then we can continue to learn in considerable detail only about the specific molecules at different points in the system and something of the topology of local patches of the control structure.

The fundamental question, then, must be what can we learn of the total control structure from studying different local segments of it ? The answer, of course, is that we can learn about constraints on the local topology which will become major clues to make good guesses about the global control structure. For example, we can ask about the average connectivity of the control topology and its variance ; does each component have many or few other components which directly affect its relevant behavior ? Classical operons seem to have two major metabolic control inputs—apo-repressor and corepressor, or inducer and repressor. Further, we can ask what kinds of function components realize on their control inputs, and with what distribution among the components. For example, an inducible operon realizes the I F function on its inducer and repressor, being inactive only if the repressor alone is present, but active for the other three input possibilities.

Based on these kinds of information, we shall have to make guesses about the overall topology of the metabolic control network. Yet for any local constraints, there will exist a very large set of possible global structures consistent with those constraints. With respect to some behavioral properties, different members of this set of structures may behave in radically different ways. Therefore this approach can work *only* if : (1) there should be, for most of the systems constructed within the constraints known from local pieces of the structure, aspects of their behaviors which are common despite diversity of structural detail. That is, there must be behaviors which are insensitive to structural detail. And, (2) those common aspects should be interpretable as aspects of the behaviors of real cells and organisms. If this should occur, then those common aspects of cells' behaviors would emerge as the explananda of a theory which would explain them to be the membership of cellular metabolic control systems in the appropriate class of 'biological like' dynamic systems. Indeed those common behaviors would emerge as the appropriate macroscopic biological observations of theoretical and developmental biology.

239

Metazoan's entire control system

The character of such an explanation warrants comment. First, it is undoubtedly too strong to insist that the behaviors to be explained be common to *all* members of the class of systems generated consistent with the currently-known local constraints. As long as the behavior in question is common to most members of the set, one can be appropriately confident that membership in the class accounts for the behavior. Second, if the biologic behaviors one is seeking to explain occur only among a (small) subset of the large set of structures and are therefore improbable with respect to the initial set of local constraints, then additional local constraints must be sought for which will restrict the class to the appropriate subset. Third, we note similarities and differences of such an explanation to quantum mechanical or statistical mechanical explanations. In these, behaviors which are virtually deterministic averages over unknowable, enormous numbers of microstates, are explained *without* assuming that the underlying microsystem remains in one microstate, but rather that it wanders ergodically among all microstates, thereby guaranteeing the class averages. By contrast, for a given organism's (unknowable) control net, that net does not alter structure ; for the species of which it is a member, their nets represent a small sample of fixed structures. The hope is that such control nets are 'typical' members of the class generated by given local constraints. If the set of all organisms are sufficiently 'ergodic' with respect to that class of nets, then statistical distributions of properties in the class of control nets consistent with local constraints are to be verified by distributions of features among all organisms. Verification of such distributions would tend to indicate ergodicity during evolution with respect to the class of systems consistent with the local constraints. Such averages have a strong similarity to explanation in statistical mechanics. On the other hand, we will wish to explain a single organism's behaviors as being typical of the class, when we cannot establish in detail its control net. Explanation of the behavior of a single organism as a typical single member of a class is not similar to statistical mechanical or quantum mechanical explanations of class averages over a system with enough components to render those averages deterministic.

It is worth noting how this form of explanation by membership in a class of dynamic systems runs very much against the grain of current molecular biology. Consider the lac operon. Here it is proving possible not only to frame a detailed hypothesis about every component and chemical step in the system, but to isolate and verify virtually all the relevant components and interactions of the model. Here we can know the entire relevant structure and can therefore deduce and verify to a high degree detailed claims about its behaviors. By contrast, when confronting the global metabolic control system, it is precisely the entire structure which we cannot reasonably know, yet we must account for behaviors which depend on that structure. Explanation by class membership of behaviors which are structurally insensitive appears to me the only possible way to do this short of knowing all the connections in detail. On the other hand, explanation by class membership, where the class is generated from local constraints on the topology, poses real problems for experimental confirmation. How shall we establish that the behaviors of cells which appear in most members of the class of systems are in fact due to membership in that class ? Knowledge of the entire control system, which might allow us to deduce the desired behaviors directly, is, by hypothesis of complexity, denied us. All we can experimentally study are local patches of the structure, distributions of the characters of local patches, and global behaviors. And given our ignorance about the detailed structure of the biological system, it may be one which does not display the common behaviors due to membership in a given class of systems but may instead be due to other factors. Experimental confirmation in the fashion of molecular biology confronts additional difficulties. The system cannot be isolated. The local constraints almost certainly cannot be manipulated. Addition or deletion of a few components describable by the same local constraints leaves the modified biological system a member of the same class of global systems and may yield no specific predictions. Perhaps the best that can be hoped for are : (1) use of the theory to suggest structure-insensitive behaviors not previously noted in organisms ; (2) use of the theory to suggest new features about the local topology.

There are additional problems in assessing any current theory which proposes to deal with global features of an entire gene control system. Any theory must utilize rules describing the behavior of a component with respect to its inputs. Such rules may be detailed and accurate or idealizations. A gene control system of 10^4–10^6 elements has so many components which realize such non-linear functions, that no current mathematical techniques, analytical or computer, allow direct study of large model systems with detailed realistic component rules. To gain any insight into the behavior of any system extensive idealization is required. For example, I treated genes as binary switches [3-5] and Goodwin created systems of neutrally stable biochemical oscillators [6]. How can we examine the range of validity of the idealizations introduced ? We need good grounds to think that the qualitative features characteristic of a class of system built consistent with known biological local constraints and with extensive idealization of component rules will remain qualitatively similar with more realistic rules. But we cannot study large systems with realistic rules in order to establish the similarities to idealized systems. We can study only the similarities of small systems with realistic and idealized components and attempt to extrapolate to large systems, or perhaps establish theories mapping behaviors of idealized to non-idealized systems. For example, Dr Leon Glass and I are trying to show that steady states in switching nets map 1 : 1 to a subset of the steady states of a homologous net whose elements realize continuous sigmoid functions on their inputs, and to gain insight into the relation between oscillations in switching and continuous nets

It is apparent that this form of explanation and confirmation will remain un-popular with the mainstream of biologists who rightly admire the precision and completeness of analysis which can be applied to small control systems. Yet it is very hard to see how, if we are to deal with very large control systems, this approach, with its limitations, can be avoided.

On the other hand, success in finding structurally insensitive behaviors with respect to a set of local constraints does avoid, for those behaviors, the epi-stemological crisis of analyzing the entire system. To state again briefly the problem posed by the example of a computer program : There is a spectrum of sensitivities of behaviors to structural modifications in complex systems. Our experimental techniques currently appear able to give us direct information only about local pieces of cellular control structures. From the constraints learned thereby we can know the global structure only to within a class of systems. Thus only properties (relatively) invariant with respect to that class can be explained by the local constraints used. Such a theory, however, would by itself remain incapable of accounting for either those behaviors which were structurally sensitive with respect to the local constraints used, or for particular connections in the structure, since by hypothesis of complexity only some components and connections would be known.

Clearly the most likely candidates for aspects of cellular behavior, which might be common features of most control systems consistent with discoverable local constraints, are behaviors which do not depend upon either the detailed molecules or detailed couplings. Such features will be likely to be among those found in *all* organisms, since we may expect reasonable diversity of control structures across all organisms, regardless of similarity of components such as DNA, RNA, and proteins, and processes like transcription and translation. On the other hand, any given set of cellular behavioral characteristics common to all or most organisms may not reflect structurally insensitive features common to the behavior of most control systems consistent with known local constraints, but instead be due to the common occurrence in most organisms of some improbable (with respect to the known local constraints) control structure, with structurally sensitive behaviors which evolution has selected, or, of course, may reflect behaviors associated with specific components (for example, DNA and coding) occurring in all organisms.

There are preliminary reasons to think a set of behaviors common to all organisms may be due precisely to general topological features common to most dynamic systems with local constraints similar to real biologic control systems.

A major constraint in cellular metabolic systems is that most components' activities are directly affected by very few other components. For example, as noted above, operons appear to have two major control inputs. Efficient auto-catalysis in organisms has evolved toward highly specific catalysis. Precisely the meaning of high molecular specificity is few control inputs per component.

As I have reported extensively elsewhere [3-5] if a component (gene, mRNA, enzyme) is idealized as being simply active or inactive, then almost any dynamic switching system constructed with few—2 or 3—inputs per element has the following properties:

(1) Permanent modes of behavior—cyclic and steady state—are restricted to exceptionally localized regions of state space, $\sqrt{N}$ for a net of N components and 2^N states;

(2) such systems have very few, $\sqrt{N}$, distinct permanent modes of behavior;

(3) the permanent modes of behavior are stable to most slight perturbations to component activities;

(4) some perturbations cause a system to flow from one to another permanent mode, but from any mode, only a few other modes of behavior can be directly reached;

(5) such systems can evolve in the sense of keeping permanent modes of behavior unaltered, but altering the flow probabilities between modes induced by perturbations.

These ubiquitous properties parallel several global properties of real cells in all organisms, which therefore are unlikely to be due to the precise control topology of any organism. These include:

(1) A stable cell type corresponds to a very restricted set of states of gene activity;

(2) the number of distinct cell types in an organism appears not only a monotonic

243

function of the complexity of the organism's DNA, but seems roughly a square-root type function of the quantity of DNA in the organism's cells ;

(3) cell types seem stable to many perturbations ;

(4) just as no mode of behavior in the theoretical systems can 'differentiate' directly into more than a few other modes, so also, throughout phylogeny, no cell type differentiates directly into more than a few other cell types ;

(5) Finally, cell types in men and fish are much the same, so the control system must allow evolution of flow paths between cell types while leaving cell types unaltered.

We noted above that attempts to confirm any such theory based on class membership face real difficulties, but suggested that deduction of unsuspected global regularities from the theory would aid. At least four unsuspected global features emerged from this theory.

(1) The number of cell types increases roughly as a square-root type function of an organism's DNA.

(2) Cyclic patterns of component activity vary in length as a square root of the number of components, and average cell cycle times also vary roughly as a square root of the quantity of a cell's DNA.

(3) The distribution of cell cycle time for cells with about the same amount of DNA is not only skewed toward short cycle times when time is plotted linearly, but *remains* skewed when time is plotted on a logarithmic scale. The same holds for the distribution of expected cycle times in random nets with the same number of components. Persistence of similar skewing on a log scale in data and theory is evidence of reasonable similarity in the underlying distributions, since many biological data which are skewed in a linear plot are normal on a logarithmic plot.

(4) The theory predicts that diverse cell types are identical in the state of activity of about 7 to 8 of their genetic loci. Berendes [7] comparing chromosome puffing patterns in stomach, salivary, and Malphigian cells of Drosophila hydei found identical patterns in about 0·85 of 110 loci. New DNA, RNA hybridization techniques, responsive to unique sequences, may soon allow testing of this prediction in other organisms.

Confirmation for the theory may also be gained by insight it sheds on unnoted characteristics of local structure. For example, the 'good' global behaviors in switching nets with few inputs *require* that many or most elements realize what I have called 'forcible' switching functions. A forcible switching function is one in which one value on at least one input line wields a veto power on that component's activity, guaranteeing that the component becomes active (or inactive) regardless of the states of the other input lines. For example, a repressible operon with two inputs—apo-repressor and corepressor is forcible on both input lines. Establishing the ubiquity of components which realize forcible functions would help confirm the theory, as would evidence for the prediction that cellular forcible components are likely to be connected into extended forcing structures.

That a first unsophisticated try, utilizing only our crudest local clue about the topology of cell control structures—that each component has few direct control inputs—should already yield a class of systems whose global behaviors are so strikingly biologically good, seems to me a major reason to be encouraged that a great deal may be learned about cellular behavior by educated guesses from local parts of cells' control systems to possible global topologies, and study of typical members of that class, without the necessity to unravel in detail the entire control structure.

It should be stressed that data on average connectivity and types of functions realized does not exhaust clues to global control structure and behavior available through study of local pieces. For example, most known enzymes exhibit monotonic activity functions with respect to their substrates in *in vivo* ranges. My colleague, S. Newman, has recently shown [8] that restriction to such monotonic continuous functions, with a few other biologically reasonable restrictions, is a sufficient condition to guarantee negative divergence in the metabolic system's vector field. Thus, behavior flow lines converge to attractor regions which jointly comprise a proper subset of state space. Here an important global result on metabolic behavior is derived from the character of local regions only.

A second example indicates another way further local data can increase the accuracy of guesses about global control structure. Suppose we learn the mean and variance of numbers of inputs and outputs per element. The members of the set of systems consistent with those constraints will have loops in them, and may be characterized by a histogram of the lengths of the shortest structural loop, if any, by which each component may influence itself. If the number of components is large, the probability of a structure with no loops, very long 'shortest' loops, and so on is very low, and can be calculated. Information about the distribution of shortest loop lengths is available through study of reasonably local pieces of cellular control structures, and can quickly be used to test whether or not further constraints need to be imposed on the local rules.

Thus, not only are the most preliminary results from the minimal assumption of low connectivity encouraging in the extent to which biologic-like dynamic properties emerge in typical class members, but the parallels may reasonably be expected to increase when other constraints, discoverable from studying only loca pieces, are included. For example, a large variance in the number of output lines per component would suggest a somewhat hierarchical global structure. Interestingly, switching nets constructed with few inputs per element, but with hierarchical structures, seem to behave identically to non-hierarchically built systems with respect to the global properties discussed earlier, but in addition now exhibit new features since some components control very many components. This example indicates that additional local constraints, restricting the class of permissible global systems to a subset of the class consistent with the initial local constraints, may often restrict to a subset in which the common behaviors of the larger initial class remain common. Thus, cellular behaviors which were explained by membership in the larger class, generated from fewer local constraints, need in no way be vitiated by the elucidation of additional local constraints.

In sum, with the incapacity to establish in detail cellular control structure, our reasonable research program must be, on the one hand, to elucidate local constraints, and on the other, to search for those properties of cellular control systems which we think will prove to be structurally insensitive global behaviors of the system. If this optimism and program proves well founded, it may be possible to explain many general features of cellular metabolic behavior without having directly to confront the epistemological catastrophe of trying to analyze in detail the control connections among all a metazoan cell's components. On the other hand, biologists accustomed to the luxury of detailed deduction of a control system's behavior and verification of all relevant parts of the proposed structure and its behavior will need to accept a form of explanation where knowledge of structure remains incomplete and deductive inference and verification is considerably less detailed. But, as Aristotle cautioned, it is foolish to demand more from a field of enquiry than it can provide.

References

1. W. M. Elsasser, *Atom and Organism, a New Approach to Theoretical Biology* (Princeton University Press 1966).
2. W. M. Elsasser, The role of individuality in biological theory, in (C. H. Waddington, ed.) *Towards a Theoretical Biology 3: Drafts* (Edinburgh University Press 1970).
3. S. A. Kauffman, Metabolic stability and epigenesis in randomly constructed genetic nets, *J. theoret. Biol. 22* (1969) 437-67.
4. S. A. Kauffman, Behavior of randomly constructed genetic nets : binary element nets Behavior of randomly constructed genetic nets : continuous element nets, in (C. H.

Waddington, ed.) *Towards a Theoretical Biology 3: Drafts* (Edinburgh University Press 1970).

5. S. A. Kauffman, in (Moscona and Monroy, eds.) *Current Topics in Developmental Biology, vol. 4* (in press).

6. B. C. Goodwin, *Temporal Organization in Cells* (Academic Press : London 1963).

7. H. Berendes, *J. exp. Zool. 162* (1966) 209.

8. S. Newman, Lecture given at the First International Meeting on Developmental Biology held at Nice, France, 1971.

Laws and constraints, symbols and languages

H. H. Pattee
State University of New York at Buffalo

In an earlier paper, 'How Does a Molecule Become a Message ?' written for the 28th Symposium of the Society for Developmental Biology [1], many of the ideas in my chapter 'The Problem of Biological Hierarchy' [2], in the third volume of *Towards a Theoretical Biology*, were further developed. I would like to add here a summary of my more recent thoughts on these fundamental problems as they have evolved since writing those papers.

In 'The Problem of Biological Hierarchy' I state as a central problem of general theoretical biology, '. . . to explain the origin and operation (including the reliability and persistence) of the hierarchical constraints which harness matter to perform coherent functions'. I might have better added, the origin, operation, and *evolution* of these constraints, since new hierarchical levels evolve repeatedly. And I conclude there that in order to find an explanation at any level, '. . . we will have to understand what we mean by a record or a language . . . ultimately in terms of physical concepts. We will have to learn how collections of matter produce their own internal descriptions.'

The difficulty is that any kind of description presupposes some form of language structure. As I conclude in 'How Does a Molecule Become a Message ?'—'A molecule becomes a message only in the context of a larger system of physical constraints which I have called a "language".' Now a language must be a closed set of structures, which we call symbols, some rules for joining and transforming these symbols which we call the grammar, as well as a set of rules for interpreting the meaning of such a collection of these symbols (for example, [3]). What has happened is that I have begun with the problem of explaining a *constraint*, which may at first sight appear to be a relatively simple physical concept, but end up trying to explain a *language* which is a very abstract concept that no one fully comprehends. It may seem that I am trying to explain the simple in terms of the complex.

I want here, in this present note, to re-emphasize my conviction that dependence on symbol structures and language constraints is the essence of life ; that it is in fact the *objective* separation of the simplified symbolic description of the organism from its detailed physical reactions, that is, the separation of genotype and phenotype, which makes the evolution of living systems so profoundly

248

different from non-living systems. Now, you will say, what is new about this idea ? The genotype and phenotype have been distinguished for well over a hundred years, and it is also well known that evolution depends on this separation. Yes, that much is true, but it is not enough to recognize the universal occurrence of this separation in living matter. To the physicist, this separation of genetic description and phenotypic function is not a trivial question, and furthermore the meaning of these concepts does not appear clearer as we discover more details. In fact, the acceptance of the structural data of molecular biology as 'the physical basis of life' tends to obscure the basic question rather than illuminate it. We are taught more and more to accept the genetic instructions as nothing but ordinary macromolecules, and to forget the integrated constraints that endow what otherwise would indeed be ordinary molecules with their symbolic properties.

What I would like to counteract is the oversimplification, or perhaps what is better called, the evasion of the genotype–phenotype distinction. In order to have an explanation of life, this distinction cannot be treated as merely a descriptive convenience for what is popularly assumed to be the underlying molecular basis of life. It is not the structure of molecules themselves, but the internal, *self-interpretation* of their structure as symbols that is the basis of life. And this interpretation is not a property of the single molecule, which is only a symbol vehicle, but a consequence of a coherent set of constraints with which they interact.

I would like to explain why the physical concept of constraint, especially in the context of origins and evolution, is by no means as simple as elementary physics textbooks lead us to believe. I would also like to suggest, though I am far from implementing this suggestion, that the concept of language need not be as complex as linguists teach us. Part of this discrepancy is simply our familiarity with what are usually the most complex language structures, such as man's spoken language, and with very simple constraints, such as a table top. But the concepts of constraint and language are very general, and closely related at a deep level. In fact, I have argued that the most fundamental concept of a constraint in physics depends on an alternative description, and that the apparent simplicity of constraints is in fact a property of the language in which it is described. On the other hand, the most elementary concept of language requires a coherent set of constraints to form its rules of grammar and interpretation. Thus we have the chicken-egg paradox in a new form : 'Which came first, the language or the constraint ?'

249

Laws and constraints, symbols and languages

▶ *Laws and constraints*. The concept of natural law in physics is quite distinct from the concept of a constraint. A natural law is inexorable and incorporeal, whereas a constraint can be accidental or arbitrary and must have some distinct physical embodiment in the form of a structure. Very often, however, especially in abstract mathematical descriptions of dynamical systems, there is no *formal* way of distinguishing the laws from the constraints. Simply writing down a system's dynamical equations of motion, for example,

$$\frac{dx_i}{dt} = f_i(x_1, \ldots, x_n, t), \quad i = 1, \ldots, n \quad ,$$

in a general system of coordinates does not reveal whether it is obeying only laws or constraints or both. If we want to predict only the local behavior of the system, this distinction may be unnecessary, but in the context of evolution, the only crucial question is how the constraints themselves came about in the course of time.

Constraints, unlike laws of nature, must be the consequence of what we call some form of material structure, such as molecules, membranes, typewriters, or table tops, these structures may be static or time-dependent, but in either case it is important to realize that they are made up of matter which at all times obeys the fundamental laws of nature in addition to behaving as a constraint. What does this mean ? If the laws of motion are complete and inexorable, what more can be said ? Why isn't an equation of constraint either redundant or inconsistent ?

The reason that constraints are not redundant or inconsistent with respect to the laws of motion is that they are *alternative descriptions* of the system. Constraints originate because of a different definition or classification of the system boundaries or system variables even though the equations of constraint may be in the same mathematical form as equations of motion. Our usual justification for choosing to use such auxiliary conditions in place of the detailed dynamics is that it *simplifies* our description of the behavior of the system. For example, the collection of molecules which make up a table top is hopelessly complex as a microscopic dynamical system, but for many practical purposes (that is, its function) it can be alternatively described as a static constraint in the form of a fixed plane, say, $z = $ constant. Fixed constraints are very useful simplifications, but it is the time-dependent constraints, especially the non-integrable constraints, which generate new levels of organization.

▶ *Constraints, rules, and regulations*. Time-dependent constraints often appear to us as embodiments of rules. Thus a very complex, dissipative, arrangement of molecules may function as a switch, in which case we often choose to ignore all the dynamical complexity and express the constraint by a simple switching rule. Complex systems may be described by many levels of constraints depending on the degree of simplification which is useful. Thus the abstract switching rule may be useful for a logical analysis of ideal networks, but much too simple for an analysis of the reliability of this function in any real computer.

It is also easy to extend the somewhat subjective concept of *rule* to its more active meaning of *regulation*, which is defined, among other things, as 'control according to a rule'. Control constraints or mechanisms are, of course, a very complicated and ill-defined set of structures. But in essence control implies that a system possesses *alternative* behaviors, and that owing to the particular nature of the constraint it is possible to correlate a controlling input variable or signal with a particular alternative output dynamics according to a rule.

Again it is important to realize that *controls must operate between different descriptive levels*, just as constraints must be defined by different descriptive levels. This is necessarily the case for all measurement, recording, classification, decision-making, and informational processes in which a number of alternatives on one level of description is reduced by some evaluative procedure at a higher level of description. Why are these necessarily two-level processes ? Why are two distinct descriptions necessary ? Because we cannot speak of an event as being both possible and impossible using the same level of description. On the lower, unconstrained level the alternatives must be possible ; for if they were impossible then deciding for or against them would be a vacuous process. But on the upper, constrained or controlled level, in so far as the rules are reliable or effective, some of these alternatives are actually selected, or more precisely, made more probable, that is, catalyzed [4]. This is one fundamental reason for the necessity of hierarchical levels of control which are characteristic of biological organization [2]. I suspect it is also at the root of the measurement problem in which the description of physical events (equations of motion) cannot be used directly for the description of the measurement of these same events [5]. This argument is also very similar to the logician's argument that any description of the truth of a proposition must be in a richer language (metalanguage) than that in which the proposition itself is stated (object language).

But now we are speaking as an outside observer who chooses his descriptions quite subjectively for the purpose of simplifying a problem he wishes to solve or for providing a simple function by a clever use of controls. Living systems on the other hand are created by their constraints and function quite independently of the biologist or physicist who studies them. How do spontaneous constraints

251

arise in matter ? Or more exactly, how do we recognize inherent constraints which have evolved autonomously, rather than from the observer's search for simplification or control ? How do we distinguish the living system's rules and functions from our subjective attempts to describe such enormously complex systems ?

▶ *Does life depend on laws or constraints* ? At this point I believe that the naive realism, characteristic of classical physics and modern biology, runs into serious difficulty when it is applied to explanations or reductions of life to physical laws. Some constraints could in principle be looked at in terms of their detailed dynamics and would be found to obey the laws of motion. This is the way we try to answer the question 'How does it work ?' when we are presented with a complicated machine or functional constraint. But as I point out in 'The Problem of Biological Hierarchy', the question, 'How did the functional constraints arise ?' cannot be answered so directly because, as we have said, the constraint is not a deterministic consequence of the detailed dynamics of the system but an *alternative description* of the whole system taken as a functional unit.

Now if this alternative description is regarded simply as the outside observer's way of handling the complexity ; that is, if the concepts of constraint, function, hierarchical levels, genotype and phenotype, and so on, are regarded as only a useful manner of speaking about the underlying physics and chemistry, then there is no objective difference between living and non-living matter. From this perspective, life appears as nothing but a very complex physical system which we as observers are forced to describe in terms of hierarchical levels of organization. Consequently, in this view the origin of life and its macroscopic evolution is regarded as only a gradual increase in complexity which has necessitated new forms of description on the part of the observing biologist.

This attitude of naive realism runs against almost all modern interpretations of physics. To regard the distinction between genotype and phenotype as based on only a kind of historical biological utility, which can now be replaced more accurately by the 'newly-discovered' underlying physics and chemistry of nucleic acid molecules and enzymes (that is, 'the physical basis of life') is very much like claiming that symbols and records can be accurately understood by a detailed physical and chemical analysis of the symbol vehicle or record structure. We know that this is not the case, and that symbol vehicles are largely arbitrary 'frozen accidents'. It is only the integrated set of rules of grammar and interpretation that gives these particular physical structures their symbolic attributes. What constitutes an 'integrated set' or a 'language' is of course the basic problem.

252

In physics, this problem arises in the concept of measurement. Measuring devices are non-integrable constraints which classify and record alternatives, and are not subject to detailed description in terms of the underlying dynamics, even in principle. Measurements are the result of the interaction of a microscopic dynamical system with a special type of dissipative constraint that can so far only be understood by its alternative, statistical description. Furthermore, physicists often regard the dissipative measurement process as more fundamental than the unobservable, formal determinism of the dynamical laws. Thus Born, for example, argues that observation itself is primary : 'for whether in a concrete case a cause-effect relation holds can only be judged by applying the laws of statistics to the observations' [6]. And Wigner in a recent review of the epistemology of quantum mechanics concludes : 'In my opinion, the restriction of quantum mechanical theory to the determination of the statistical correlations between subsequent observations reproduces most naturally the spirit of that theory' [7].

In any case, whether one regards the laws or the measuring constraints as primary, it is well known among physicists (and should be better known among biologists) that a clear and unified description of events and records of events has not yet proved possible in quantum theory. It is for this reason that constraints in physics need not be so simple ; and in fact it is precisely for those non-integrable constraints which are necessarily associated with the writing and reading of symbols—whether for genetic records, controls, or measurements— that unity and objectivity is lacking. Furthermore, if this unity is lacking between events and records of events in physics, then it is not easy to understand how, through the facts of molecular biology, this unification can appear.

On the other hand, by the study of *evolutionary theories*, in particular, theories of the spontaneous origin of life and its hierarchical levels of control, I think we find clues to the solution or at least to the difficulties of finding an objective basis for the separation of matter and symbols. What are some of these clues ?

▶ *Self-constraint and self-rule*. It is my central idea or strategy that *the essence of the matter–symbol problem and hence the measurement or recording problem must appear at the origin of life* where the separation of genotype and phenotype through language structures took place in the most elementary form. Studying the problem in this context helps remove the physicist and logician from the measurement problem. At least it helps widen his tacit anthropocentric assumptions about the function of measurement processes, shifting the emphasis from his brain to the environment or ecosystem where selection takes place. Furthermore, at the level where life originated, the problem must appear greatly reduced in complexity, although it may well be necessary to generalize our concept of records, symbols, and languages to apply them at that primeval level. The relation of this approach to the more conventional physical study of the quantum theory of measurement I have discussed in 'Can Life Explain Quantum Mechanics ?' [5].

The necessity of an objective criterion for the occurrence of a separation of events and records has also led me to the hypothesis that *the constraints of living matter must contain their own descriptions*. This follows from the physicist's concept of a constraint as an alternative description of an underlying dynamical process, which suffers, as we said earlier, from a basic subjectivity. How do we know that a constraint is nothing more than a convenience for the higher purposes of human computation or control ? The only answer I have found is in objectifying the description itself. But of course this shifts the problem to the objective description of the description which at first sight sounds like the beginning of an infinite regress. It is here that we must turn to the fundamental property of a language which, as we said at the beginning, is itself a larger, coherent system of constraints. How does this help evade the infinite regress ? If one constraint gets us into this difficulty, how can adding more than one get us out ?

This same problem is stated for a language in the following form : How is one symbol or word given a definite meaning ? By a coherent set of other words which we call a definition. Then how are the words of the definition given meaning ? This also sounds like an infinite regress, but we know that in language this problem is solved. There exist many finite sets of words which can not only define themselves, so to speak, but also define the grammar, as well as form meaningful statements about symbols or groups of symbols, which are called metalinguistic statements. A language therefore possesses the property of self-description or, in the more physical terminology, self-constraint.

There remain, of course, many fundamental questions about language. What is the simplest self-constraining set ? Can simpler sets be expected to evolve this self-referent property ? How do the grammars of such sets evolve ? This type of question tends to emphasize the abstract, symbolic aspects of language, but what I believe more important for biology are the physical properties which make possible symbolic behavior in the first place. Even the most abstract symbols must have a physical embodiment, however arbitrary the symbol vehicle structure may be. Instead of requiring simply a finite, self-*defining* system in the abstract symbolic sense, it is more fundamental to require a finite, self-*constructing* system in the direct physical sense. This implies a set of constraints which in some coordinated way can reconstruct themselves, as well as establish rules by which other structures can be generated. This coordinated set of constraints would amount to a language structure, that creates a new hierarchical level of organization by allowing alternative descriptions of the underlying detailed behavior. But the problem of the spontaneous origin of such a set remains.

▶ *The origin problem.* There are two very general pictures that one may form of how coordinated sets of constraints might arise spontaneously from more or less chaotic beginnings. The most common picture is that of elementary units assembling themselves into larger units. For example, the origin of life is commonly pictured as starting with simple molecules of carbon dioxide, methane, ammonia, and water, and through the activation of some energy source gradually building up amino acids, sugars, bases, then polypeptides and polynucleotides, and finally cell-like aggregations of these macromolecules, until eventually the minimum complexity of living cells is finally reached from which biological evolution can proceed. Some stages of this picture have already been demonstrated in abiogenic experiments. This sequence of origin might be called the formation of *complexity from simplicity.*

A more subtle and less easily pictured process of spontaneous organization can come about the other way around. We imagine initially chaotic aggregations of extreme complexity within which there arise persistent regularities which, so to speak, condense out simple behavior. This formation of *simplicity from complexity* is inherently a collective or global activity of the entire aggregation, as opposed to the locally specific nature of organization created by aggregation of special structures. It is this latter picture of spontaneous organization that accounts more easily for the origin of new hierarchical levels of control, which includes integrated sets of constraints that I have called language structures. For example, while specific enzyme-like polypeptides might arise by the spontaneous-assembly process of our first picture, it is unlikely that this same process could account for the genetic coding enzymes which require an integrated set of highly cooperative constraints.

Unfortunately, specific origins of the second type are much more difficult to model or demonstrate experimentally since they depend on the detailed properties of a globally complex system. On the other hand, the general type of simplification from complexity has been illustrated by models, for example, Thom's catastrophes in his topological dynamics [8], dissipative structures in non-equilibrium thermodynamics (for example, [9]), and Levins' spontaneously simplifying complex systems [10]. The behavior of Kauffman's randomly

255

connected, random switching nets also illustrates this behavior in fixed sequential systems [11].

These models and theories are of great help in biology because of their generality and applicability to the problem of the internal generation of constraints. However, they lack many of the known physical and chemical interactions which in my opinion lead to the formation of the most significant constraints. For example, it is hard for me to imagine how the origin of selective, catalytic growth characteristic of present cells can be modeled without including some representation of selective monomer addition steps in individual growing copolymer molecules. In other words, the system must generate internal constraints at the molecular level as well as at the statistical or macroscopic level.

On the other hand, all the demonstrations of abiogenic syntheses at the chemical level, while of great significance for verifying the complexity-from-simplicity theories, lack a realistically complex macroscopic environment. Since the initial conditions of these experiments are kept as simple as possible, they are unlikely to generate constraints of the simplicity-from-complexity type.

▶ *Can we test origin theories ?* How can we design experiments which could demonstrate the spontaneous appearance of constraints which internally simplify the behavior of a complex system ? How do we recognize such simple behavior as inherently generated by objective constraints rather than by the outside observer's subjective classifications of the system's behavior ?

The first requirement is that the choice of initial conditions be guided by realistic appraisal of the complexity of any primitive earth environment, and not by pre-experimental selection on the part of the designer of what is presumed to be significant. We should think of the initial problem as one of accurately simulating a primitive, sterile ecosystem, rather than biasing our analysis in favor of specific reactions or products. I have for some time suggested this type of simulation as a complement to the many abiogenic synthesis demonstrations [12-15]. My first choice would be a simulated sterile seashore with primitive atmosphere, diurnal radiation, sand and clay, waves and tides, and so on, all of which could not have been reasonably missing from the primeval seashore, and all of which seem likely to have significant effects on the chemical development of almost all the reactants, especially any copolymers. Our initial observations of such a system must be more in the style of a naturalist rather than that of a biochemist. But as regularities or predominant behavior appear we must determine if specific catalytic, structural, or thermodynamic constraints are responsible. By their very nature, such simulations are large and expensive and will require cooperative effort from several fields for their design and monitoring. On the other hand, compared to the size and cost of modern exobiology experiments in the form of planetary explorations, such terrestrial simulations are relatively inexpensive and can be expected to provide essential data that could not be obtained any other way.

Even if such realistically complex simulations should develop internal, co-ordinated constraints relevant to origin problems, this is not equivalent to a theory of origins. A theory need not provide detailed predictions, but at least one might hope that from the theory it would be possible to eliminate unessential complications which may occur in primitive earth simulations. In other words, one would hope to apply the theory to computer programs which might then evolve new hierarchical levels of organization. One difficulty is that all present programming languages operate in a purely symbolic, sequential mode with no necessary physical constraints. As Conrad has suggested in this volume [16], the cost of achieving algorithmic prescription of sequential operations may be the loss of precisely those global condensations which are essential for self-constraint or self-simplification. Whatever the case, we shall learn more about origins only by serious theoretical and experimental study of the problem—the same course that we follow in other sciences.

References

1. H. Pattee, How does a molecule become a message? in *Developmental Biol. Suppl. 3* (1970) 1-16.

2. H. Pattee, The problem of biological hierarchy, in (C. H. Waddington, ed.) *Towards a Theoretical Biology 3: Drafts* p. 117 (Edinburgh University Press 1970).

3. Z. Harris, *Mathematical Structures of Language* (Wiley : New York 1968).

4. H. Pattee, Physical problems of decision-making constraints, *Int. J. Neuroscience 3* (1972) 99.

5. H. Pattee, Can life explain quantum mechanics? in (T. Bastin, ed.) *Quantum Theory and Beyond* p. 307 (Cambridge University Press 1971).

6. M. Born, *The Natural Philosophy of Cause and Chance* p. 47 (Dover Pub. : 1964).

7. E. P. Wigner, Observations at the end of the conference, preprint for *On the Epistemology of Quantum Mechanics* (London, Ontario, March 1971).

8. R. Thom, Topological models in biology, in (C. H. Waddington, ed.) *Towards a Theoretical Biology 3: Drafts* p. 89 (Edinburgh University Press 1970).

9. I. Prigogine, R. Lefever, A. Goldbetter and M. Herschkowitz-Kauffman, Symmetry breaking instabilities in biological systems. *Nature 223* (1969) 913.

10. R. Levins, Complex systems, in (C. H. Waddington, ed.) *Towards a Theoretical Biology 3: Drafts* p. 73 (Edinburgh University Press 1970).

11. S. Kauffman, Behaviour of randomly constructed genetic nets : binary element nets

Laws and constraints, symbols and languages

(p. 18) continuous element nets (p. 38), in (C. H. Waddington, ed.) *Towards a Theoretical Biology 3: Drafts* (Edinburgh University Press 1970).

12. H. Pattee, in (S. Fox, ed.) *The Origin of Prebiological Systems* p. 400 (Academic Press : New York 1965).

13. H. Pattee, The recognition of description and function in chemical reaction networks, in (R. Buvet and C. Ponnamperuma, eds.) *Chemical Evolution and the Origin of Life* (North Holland : Amsterdam 1971).

14. H. Pattee, Physical problems of the origin of natural controls, in (A. Locker, ed.) *Biogenesis. Evolution and Homeostasis.* (Springer-Verlag : Heidelberg and New York. In press).

15. H. Pattee, Physical theories of biological coordination, *Quart. Rev. Biophysics 3* (1971) 255.

16. M. Conrad, The importance of molecular hierarchy in information processing, in (C. H. Waddington, ed.) *Towards a Theoretical Biology 4: Essays* pp. 222-8 (Edinburgh University Press 1972).

Biology and meaning

B. C. Goodwin
University of Sussex

This essay is concerned primarily with the relationship between knowledge and meaning. Nowadays these domains of apprehension are often experienced as antithetical, closely connected with the apparently incompatible domains of fact and value. Opposition can clarify by sharpening distinctions, but when dual concepts become irreconcilably separated, the mind has fallen into one of its own traps and the creative tension of opposites is lost. I believe that this has happened in contemporary thought with respect to knowledge and meaning, or fact and value, with unfortunate consequences regarding the development of a theory of organism. Organisms work on the basis of knowledge of their component parts or activities and of their environment. They use this knowledge to generate more or less integrated patterns of behaviour, thus creating purposeful wholes out of parts. The use of such language in the description of biological organization is hardly new, having been well developed in Aristotle's thought. It is often now regarded as mere metaphor, or worse, misleading metaphor, because of its teleological flavour. In the sixteenth and seventeenth centuries there was good reason for this criticism, but we ought to have grown beyond this now to another, though connected, interpretation of purpose and its relationship to knowledge.

In the present essay I make no attempt to formalize an approach to biological order and behaviour in terms of knowledge-using systems, ones which carry and use representations of environment and self, thus achieving a particular type of stability, relationship, and meaning. There are several detailed and very interesting problems which need to be clarified before any such attempt can be made, such as an analysis of the principles involved in making wholes out of parts ; the particular way in which hypotheses operate in determining which whole is generated from ambiguous or incomplete knowledge about parts ; what constitutes a representation or hypothesis ; and others about most of which I am ignorant. The approach I have used is to look for some historical threads of the dichotomy that has resulted in the contemporary loss of meaning, which led Eliot to ask

Where is the wisdom we have lost in knowledge ?
Where is the knowledge we have lost in information ?

Biology and meaning

One such thread runs through alchemy ; and this in turn connects with elements
in Greek and Hebrew conceptions of reality and purpose. Since I believe that
scientists today can no longer pretend to be other than philosophers and his-
torians, though usually rather bad ones, as will become evident, I make no
apologies for incursions into other domains. What I do apologize for is the
sketchiness of the treatment, its dilettante flavour. This emerges mainly as asser-
tive simplification. My objective is to examine certain assumptions about the
nature and purpose of the scientific enterprise, assumptions which I believe to
have now lost their usefulness and validity and which therefore need to be
replaced. One such is the concept of objectivity. By defining knowledge in terms
of objectivity we have lost the concept of meaning, and as a consequence science
has become amoral, unconnected with value. At the same time a science of
objects, objective knowledge, cuts itself off from an adequate theory of organism.
I am concerned with both of these problems, but at present more with the former
than the latter.

▶ *The conflict of opposites*. Two inter-twined components of human endeavour
lie at the centre of the dilemma that I would like to consider. In a philosophical
context, these aspects of mind give rise to the epistemological problem on the
one hand and the moral problem on the other. In more immediate terms, we may
refer to them as the domains of knowledge and of meaning. Our heritage splits
cleanly in two on these domains : the Greeks were very strong on knowledge,
but weak on meaning ; while the Hebrews had little time for epistemology, but
filled the world with such a vision and sense of meaning that it is only now growing dim. Of course these aspects of mind are never separated from one another.
But one can grow at the expense of the other, driving it underground, stifling it,
as has now happened in our over-development of the Greek rationalist tradition.
The endless human struggle is to keep them in balance and harmony so that they
strengthen one another. Because of the extreme polarity inherent in our heritage,
we have been presented with the dilemma in a particularly acute form. This
conflict of opposites has resulted in the liberation of quite extraordinary amounts
of physical and psychical energy in the West, a restlessness reflecting the failure
to resolve and harmonize the strands of mental activity which has carried us with
startling speed into a new age. This new age has dawned primarily as a result of
the development of science. Since this age, with its new image of ourselves and
the world, has grown out of a conflict, one might expect that the seeds of the
resolution would be inherent in the present situation, the synthesis emerging
from the dialectical confrontation of the opposites. I believe this to be the case,

260

B. C. Goodwin

but it is of course exceedingly difficult to pick out relevant contemporary movements. That there is a general and growing disillusion with science and technology is evident ; but where the resolution lies is not clear. It is easier to look for
some historical antecedents and parallels, possibly picking up a thread that could
help to unravel the present tangle and so release once again man's creative
potential in a new synthesis of knowledge and meaning, a new understanding
and vision.

▶ *The alchemical synthesis.* The application of knowledge leads to power ; the
application of meaning leads to wisdom. One of these without the other is always
dangerous. Use of power without wisdom is such a familiar abuse that little need
be said about it. We are at the beginning of a debate on this issue in relation to
the conflict between technological innovation and preservation of the natural
order. Wisdom without power, on the other hand, is so unfamiliar to us that it
often seems eminently desirable, and indeed it is enthusiastically embraced by
the social out-groups who have renounced Western political and social institu-
tions, the hippies being the most dramatic example. Total renunciation of power
and absorption into a state of permanent vision, contemplation, and experience
of meaning, is the image we have of Eastern, particularly Buddhist, wisdom.
Believers in the Western way of life condemn this as irresponsible ; and so, I
believe, would the Jewish prophets, absorbed as they were with God's role in
history as law-giver and the problems of individual and social morality. Solomon
was wise because he applied the Law with insight and understanding. He would
not have been considered wise if he had not applied his vision of meaning to the
conduct of human affairs.

Essentially the same tension between power and wisdom is present in the
alchemical tradition. The alchemist believed that the meaning of life was hidden,
but could be perceived and experienced if one adopted the right attitude of mind
and performed the right actions. The most singular and interesting aspect of this
art is that the actions which the practitioner was called upon to perform were not
simply moral exercises and ethical taboos such as one finds in most mystical
cults ; the alchemist was required to bring about transformation of the world, of
matter and spirit, at the same time as he underwent moral transformation himself.
He was called upon to participate in the changes of substance that took place
in his retort ; indeed, it was stated explicitly that unless he so participated in
transformation, becoming himself more refined and pure as did his material, from
oxide of mercury to gleaming silver metal and, ultimately, to gold, then the
physical changes would not themselves take place. A passage in the *Musaeum*

261

Biology and meaning

Hermeticum [1] states that 'the mind must be in harmony with the work'. Another passage has it that to acquire the 'golden understanding one must keep the eyes of the mind and soul well open, observing and contemplating by means of that inner light which God has lit in nature and in our hearts from the beginning'. The essence of the alchemical process is thus quite clearly a two-way relationship between the adept and nature, both undergoing transformation as occurs in a true dialogue. The basis of this relationship was the belief that Nature has an innate tendency to seek a state of perfection, matter transforming into immortal, imperishable gold. Likewise man has a longing for perfection. He can thus learn from Nature and at the same time assist her in her striving.

The contemporary image of the alchemist as one who was concerned solely with the use of magical formulae to turn lead, copper, iron, and other lower metals into gold, thus achieving riches and power, is a materialist travesty of the most extraordinary proportions. It is roughly equivalent to the view that the pinnacle of achievement of the scientific tradition was the development of the atomic bomb, the transformation of matter into energy, and its use in the destruction of Nagasaki and Hiroshima. The current misconception of alchemy is due to the fact that it was vigorously discredited as heretical by the Church throughout the Middle Ages and the Renaissance because, like all mystical cults, it believed that God should be experienced, not simply believed in by the faithful ; and because of its own ultimate failure to resolve the tensions between knowledge and meaning, the tradition foundering under an accretion of ambiguous symbolism and virtually impenetrable jargon. Once a tradition fails, dies, and is discredited, it takes some devoted scholarship to rediscover its sources and significance. This has been provided in our time largely by the efforts and insights of C.G.Jung, who was led to the meaning of alchemical allegories via the analysis of dreams recorded by his patients, and through a comparative study of mythological and religious symbolism. His books *Psychology and Alchemy* and *Mysterium Coniunctionis* [2, 3] are masterpieces of inspired scholarship. Balanced accounts of the alchemical tradition can now be found in other volumes as well, such as Kurt Seligmann's *History of Magic* [4]. There is also a great deal of insight into the significance of Oriental alchemy and its influence in Eastern and Western thought in Needham's superb volumes on *Science and Civilisation in China*.

The aspect of alchemy that is so central to the present enquiry is the fact that this tradition attempted to fuse knowledge and meaning by combining science (*scienta*, knowledge), the study of natural process, with morality, man's attempt to realize his own perfectibility and self-fulfilment, itself a continuous process.

262

B. C. Goodwin

The difficulties inherent in such a fusion are legion. The use of alchemical knowledge without its transformational moral meaning resulted in the magical manipulation of nature for purposes of power, with the inevitable consequences of such
perversion as described in the Faust legend. The pursuit of meaning, the experience of moral fulfilment, without application to the world and hence without
the responsibility and compassion we recognize as manifestations of wisdom,
resulted in total withdrawal from the human condition and a proliferation of
elaborate, impenetrable symbolism and verbiage. Contemporary science has
essentially chosen the course of knowledge and power, split off from meaning
and morality because knowledge has become an end in itself. Scientists today
do not expect to be morally transformed by their activities. They do not, in fact,
participate in a relationship with the world that acknowledges the autonomy and
the ultimate inviolability of natural processes, the condition for a dialogue with
Nature, which has now become something to be penetrated, known, and controlled. Participation in a process of mutual transformation is in fact expressly
ruled out by the contemporary ideal of objective observation, preferably by a
machine which cannot change its state except in response to the particular
events which it is designed to record. The knowledge so obtained is regarded
as neutral, without moral 'contaminents'. It can be used beneficently or mal-
evolently, but it is always used to exercise control over the world, certainly not to
transform oneself. Thus science as we know it today has largely opted to pursue
the course of manipulation and power, drawing us inevitably into a Faustian
crisis which arises from the irreconcilability of manipulation and wisdom. To
manipulate wisely one must be wiser than Nature, wiser than Man, for both must
be manipulated ; hence one must be God. Faust found that only the Devil would
play this game with him, tempt this hubris. The corruption arises with the
decision to manipulate rather than to engage in a dialogue, the decision to be
Master rather than partner.

▶ *The scientific heresy.* What, then, is the alternative to the dominant contemporary attitude in science ? How can one redefine the scientific enterprise so
that it is coupled with the search for meaning and wisdom, not just the acquisi-
tion of knowledge and power ? In what way is it possible for us to participate in
this enterprise such that we may be transformed morally in the process of learning
about the world ? In seeking tentative answers to these questions it is instructive to try to follow the development of the essentially heretical concept of
the objective observer, the amoral, non-participating agent, to its point of self-
impalement on its own inconsistencies.

Biology and meaning

There is in fact a remnant of the ethical attitude in science which persists in the form of a commitment to the search for truth. However, this truth has become so refined in its meaning and significance, so intellectualized and divided from experience, that its moral force has been severely attenuated. To the alchemist the truth was revealed when he had achieved a union of soul and mind, of meaning and knowledge, and its symbol was gold, living gold, which he experienced within himself. Only then would he see in his crucible, the Hermetic vessel, gold emerging from the impure ingredients of the starting material as a sign that his own transformation had been achieved. There is no doubt that hallucinatory states and visions accompanied these experiences of transformation, visions that would be generally discounted now as highly subjective phenomena having not only nothing to contribute to scientific insight, but definitely detracting from it. And it is certainly true that the domain of trance and hallucination is dangerous in the extreme, bordering on total dissociation, a breaking of the dialogue. The proliferation of elaborate allegorical and symbolic descriptions of psychical states and the extreme confusion and deliberate mystification found in many alchemical texts is evidence that the Hermetic tradition was losing its way in the psychological jungle, was not always finding the universal symbols required for the communication of meaning. Despite some very penetrating insights into the structure of psychical inner space, documented in detail by Jung in his volumes, alchemy was ultimately overwhelmed by the richness and inherent difficulty of its own subject matter.

Empirical science in the seventeenth century turned its back firmly on this most elusive domain of enquiry, regarding the insights as phantasmagoria and repudiating them as being without value in the exploration of the world of 'fact' and 'reality'. Galileo's science made extensive use of observation and experimental test, providing criteria of universality that must have burst like a revelation upon minds struggling with the interminable ramblings of Medieval scholasticism or the obscurities of alchemical allegories. There seemed to be nothing firm in these voyages of the mind, no 'objectivity'. Lacking adequate procedures of classification and analysis, the exploration of inner space, of the psychological process, was first abandoned and then deliberately excluded as a domain of scientific enquiry by the Galilean tradition which eliminated qualities from its world of investigation, retaining only quantifiable observables. Man's role in scientific discovery became that of the pure observer, distilling the truth in ever more refined form from his observations and measurements. Providing the correct procedure was followed, this truth would be automatically revealed to man ; he played no creative part in the process. That is to say, what was revealed to the scientist was not just human truth, but universal truth. The same truth would be revealed to any other intelligence. Man could thus in effect step right outside of nature and become the perfect observer, not interfering in any way with natural process, much less participate in it. In the past, only God was considered to be in a position to do this if he chose. Small wonder the scientist felt himself to be in a rather powerful position.

▶ *The observer observed*. This situation in science has now broken down for two reasons. The first has arisen within science itself. Observation *does* perturb microphysical states, because observation is perturbation. It is negligible only in macrophysical processes, where the mass and energy of the observed system is very much greater than that of the observing wave packet. This recognition gives us the Uncertainty Principle, satisfactory as far as it goes.

But there is within physics the unresolved problem of how to describe the process of measurement itself : how microstates 'collapse' into unique, singular events in the macro-world. Classical or macrophysics did manage to describe the process of measurement in terms of a well-defined observer ; but the situation in physics today is thoroughly unsatisfactory, with an arbitrary and so far unbridged gap between the micro and the macro worlds.

Thus the postulate of the perfect observer has broken down within physics itself. But it has been under severe attack also in studies on the philosophy and psychology of science. The very notion of scientific truth as a unitary concept is disappearing. There are points of view and theories arising from them, defined by paradigms [5]. Different points of view are in conflict at any one time, as illustrated for example in Kearney's recent study of the scientific revolution in the sixteenth and seventeenth centuries [6]. Sometimes one view will dominate and eliminate another ; or they may become fused in a unitary, higher-order theory, such as the wave-particle duality of modern physics. In the light of this dawning self-consciousness, the scientific enterprise may be about to enter an era of altogether greater sophistication, versatility, and creativity, with scientists becoming aware of the metaphysical assumptions which underly the viewpoint they are holding at any moment in their development as scientists and as human beings. A similar rather dramatic increase in consciousness has occurred in the Arts only in this century, with the development of what Malraux has called the museum without walls [7], the work of the art historian, which allows us to see works of art as a part of the panoramic unfolding of man's history. The Lascaux cave paintings or the Ashanti woodcarving can now be seen and experienced with every bit as much validity and meaning as work by Moore or Picasso. Science is about to cross this threshold as well, to be seen and understood in the context of cultural and intellectual history, as Collingwood predicted [8]. The focus here is man's dialogue with his fellow men and with Nature, his attempt to understand himself and his world.

The subjective element can thus return to science, but within a higher level of consciousness, the scientist becoming aware of the role he is playing in presenting or supporting one viewpoint rather than another. The objection to subjective factors in scientific work in the past has always been based upon the assumption that they were unconscious factors, hence distorting, irrational ; that is, they were unconscious projections. But if one is aware of one's viewpoint and the way it is moulding one's thinking and theorizing about a particular problem, this objection is largely overcome. Furthermore, this awareness allows one to transcend the limitations and constraints that accompany the belief in 'objective truth'. This belief can now be seen as a particular metaphysical commitment based upon a set of assumptions about the nature of observation and measurement, assumptions which themselves lead to unresolved paradoxes in physics. Axioms which lead to inconsistencies cease to be convincing or compelling. The recognition that the human being plays a central role in constructing an image of himself and his world and that these images or internal models or sets of hypotheses are conditioned both by internal (psychical) and by external factors provides a degree of freedom which could liberate human creative potential in a dramatic manner, as has occurred in the arts with the dropping of the representational norm, while simultaneously engaging his ethical responsibilities in relation to this freedom. At the same time this recognition might lead to the resolution of paradoxes which have arisen from the adoption of unnecessary assumptions about the nature of scientific investigation.

These must seem rather grandiose claims and need more detailed elaboration than can be provided in this essay. The contention is twofold : (1) the recognition of man as a creative agent who participates positively and with a degree of freedom in the construction of images of himself and the world imposes a responsibility upon him with respect to his reasons for choosing one image rather than another, and can thus lead to moral involvement in the scientific process ; and (2) the realization that Man the Observer is in fact Man the Selector, seeing what he is prepared and able to see, conditioned as he is by his viewpoint, his hypotheses, eliminates the concept of the objective observer and so may lead to a more satisfactory account of the measurement process. The

system observes what it is prepared to observe, in the same way that if one constructs a trap for a particle one catches a particle ; if for a wave, one catches
a wave. This is by no means an original suggestion, having been made and
elaborated in some detail by Eddington many years ago, and developed more
recently by Bastin and Kilmister [9, 10] and others. I am not in the least competent to develop this argument, which has not yet succeeded in resolving the
difficulties inherent in the relation between quantum theory and macrophysics.
However, in a biological context, where I am somewhat more competent to
assess the possible transforming power of new attitudes, the view of the organism
as an hypothesis-generating and testing system is potentially very powerful. In
the study of artificial intelligence there is currently an active investigation of the
exact meaning which is to be assigned to the proposal that a primary function of
Mind is the generation of images, models, hypotheses about its environment, and
the analysis of its consequences in epistemology and metaphysics. The exploration of the implications of looking upon all organisms as hypothesis testing
systems has not yet, however, begun. It could transform biology by placing
model construction and observation at the centre of the biological process, not
at the evolutionary periphery, the phenomenon of Mind.

▶ *Organism and hypothesis*. Let me elaborate this approach by a consideration of
some of its consequences. I am particularly concerned to explore the possibility
of resolving some classical antinomies and their modern counterparts by means
of a change of context. One of the most celebrated of these conflicts was that
between Aristotle and Hippocrates, an atomist, on the nature of the formative
influences which are responsible for the generation of characteristic structure and
behaviour in organisms. Hippocrates contended that the general similarity of
static and dynamic form between parents and offspring was due to the transmission of specific substances from parental organs, carried by the seminal fluid,
to the embryo. His position was clear, straightforward, and entirely materialistic :
specific structure is transmitted by specific substances. The modern version of
this theory is equally clear and unambiguous : the resemblance between parents
and offspring arises in consequence of the transmission of specific genes, the
hereditary units, from one generation to the next. It is the dominant view in
biology today and preserves in remarkably pure form the traditional position of
the Greek Atomists. It appears so self-evident to the modern scientific mind that
the difficulties of this position are often not even considered ; or else these difficulties are dismissed in somewhat cavalier fashion by means of the additional
principle that all one needs to add to it is the concept of interaction between the

units, the genes, to complete the picture. The fact that the same position in physics gave rise to the still unsolved many-body problem is not regarded as a warning of limitation ; it is simply a hurdle to be jumped when necessary, with the help of a rather large computer, presumably.

Aristotle's objections to the atomist position are both simple and cogent, but his own explanation presented some equally difficult problems. How, asked Aristotle, could a young man without a beard transmit to a son the substance required to grow a beard if that substance had to come from the beard itself ? Or how could a man who had lost his hands in battle, say, nevertheless transmit the capacity to form hands in his offspring ? The problem here is that between potential generative capacity and the realization of that potential. What substance can carry such potential ? Aristotle contended that from substance alone one cannot make deductions about form ; that knowing the composition of some-thing is not sufficient to determine its structure, a position he inherited from Pythagoras. One must add to substance a principle of organization, which for Aristotle was a form or an idea, immanent in the process whereby order of a characteristic type emerges from disorder or lower order, as the embryo from the egg. Plato considered such ideas to be transcendental and autonomous, but for Aristotle the forms were also sources of energy in matter striving to organize it in some perfect order which was the final goal, the telos, of the process.

What answer do we have today to this problem of the emergence of order, particularly that of creative, novel, order ? As I have said, contemporary atomists ascribe the emergence of order to some principle of interaction between the parts of which the entity is made. This is a perfectly consistent and satisfactory idea as far as it goes. But so far it has yielded adequate solutions to simple interactions only, such as two-body problems, for example, the motion of a planet around the sun ; or to statistical regularities which arise from weak, uniform interactions between very many units, such as the ideal gas laws. This hardly seems a promising approach to the intricate organization of living systems, let alone their creative behaviour. But Aristotle fails us in this respect as well, for we want to know more about his forms or ideas. How are we to understand their nature and their operation ?

To suggest that an answer to this problem may be inspired by studies on the nature and origin of scientific theories such as those of Kuhn may seem some-what far-fetched, but let us consider where such a path may lead us. Suppose we regard living organisms as systems which generate and test hypotheses about their environment. At the psychological level this has fairly direct intuitive meaning. But what would it mean about a bacterium ? It would mean that such a system has a set of hypotheses which are present in coded, symbolic form in order to satisfy our intuition about the nature of hypotheses ; that these must be subject to variation and test in relation to an external world ; and that there must be some principle of correspondence whereby 'good' hypotheses about this world can be retained and stored while 'bad' ones are discarded. It is probably evident to the reader that the obvious candidates for such hypotheses are the genes, or more generally the hereditary material of the bacterium.

This genetic information is in symbolic form ; that is, it cannot interact directly with the external 'world' of nutrients, salts, oxygen, and so on, which define the environment of the bacterium. It has to be translated into another language, generally that of proteins, before interaction and testing can occur. A bacterium carrying the genetic information for the metabolic ultilization of the sugar lactose, known as the lactose operon, may be said to carry the hypothesis that lactose is likely to be encountered in its world, and that this substance can be transformed in a particular way to provide energy and building blocks for growth. At a more complicated level, we may say that in the chromosomes of the egg there are hypotheses about the appropriate response of a cell in the ectoderm which is required to transform into part of the lens of the eye when it encounters a cell in the underlying optic cup.

Evidently by using the language of hypotheses and hypothesis-testing I am not providing any kind of an explanation for the phenomena under consideration. I am simply redescribing them, in a different context. However, the potential value of a new context is that it may virtually do away with old antinomies. It will certainly introduce new ones, but if the context is of value, these new dilemmas will occur at a higher level than the old. What possible value is there in regarding organisms as hypothesis-generating and testing systems ? There are three consequences which occur to me immediately. The first is a suggested resolution of the atomist-idealist conflict. Hippocrates and Aristotle were both partly right, as they must have been for this argument to survive so long. Hippocrates and the atomists were correct to maintain that there are distinct substances which are transmitted from generation to generation and that the composition of the system is a very important aspect of its structure. Aristotle was correct to insist that something like formative 'ideas', different in some sense from ordinary physical matter, must guide the intricate and extraordinarily varied formative processes of organic nature. We now recognize this difference to lie in the symbolic nature of the genetic code and the remarkably elaborate system

269

Biology and meaning

cells have for its translation. It is genetic symbolism that enables living matter to step outside of the constraints imposed by physical laws formulated in terms of minimal potential criteria. The primary structure of a protein is not, for example, determined by a condition of minimum free energy. It is determined by the sequence of bases in its corresponding gene or cistron. And this is determined by some organismic criterion of successful performance of the protein in interaction with its environment. The symbolic nature of the genetic material is what provides a virtually inexhaustible reservoir of potential genetic states for evolution, since symbols can be juxtaposed in very many different ways to provide new 'statements', new hypotheses, which can then be tested.

The second obvious result of looking at organisms in this way is the possibility of transcending a conflict that has arisen regarding the importance of genes in determining a process such as embryonic development. That the hereditary material is of the greatest significance no-one will dispute. But it has been claimed that knowing the information in the genes is equivalent to knowing the embryological process, for example. Considering the hereditary material as a set of hypotheses immediately puts the genes in their proper place. For it is evident that an hypothesis is always considerably less than a total set of instructions for its interpretation and test, without which the hypothesis is useless. If a person says that he holds the hypothesis that there is life on Mars, then he is assuming a very great deal of knowledge on the part of the person to whom he is speaking. Quite apart from the basic assumption that the hearer is competent to interpret the sentence grammatically, extensive experience of the world is also taken for granted : that there is an actual entity located in a defined position known as Mars, that the hypothesis has some meaning, for example, that there is some possible relationship between life and Mars, that there are ways of testing this meaning, and so on. In like manner, the gene in the lactose operon which codes for the particular enzyme known as the lactose permease is taking for granted a great deal about the pre-existent structure and activity of the cell. If there were no cell surface, for example, no membrane, the genetic 'statement' would be meaningless, uninterpretable. So the hereditary material speaks to a competent hearer, and assumes that it can interpret and test the hypotheses it is presenting in some manner which will establish whether or not they have meaning.

And thirdly, this context provides a natural and immediate framework for comparison of different levels of organization of the biological process. It is intuitively evident that, at the level of mind, hypotheses are subject to continuous alterations as a result of experience, a testing of their relevance and meaning. In

270

contrast, within the lifespan of any single organism, genetic hypotheses cannot, according to current theory, be so altered or improved. An organism inherits a fixed set of hypotheses and its life may be viewed as an unfolding of strategies consistent with these constraints. The genetic hypotheses are subject to alteration from generation to generation, of course, by a process which neo-Darwinian theory ascribes to random mutation. Such alterations are not considered to be directed in any way, in contrast to mental hypotheses which are subject to criteria of improvement. Behaviour involving the cognitive processes can in fact be viewed as the continuous formulation, testing, and improvement of hypotheses according to criteria of behavioural success, whatever these may be. This context thus provides a natural framework within which to compare and contrast different levels of biological organization. There are finer comparisons than those mentioned, for example, between levels of behaviour involving language, a distinctly human domain, and those involving cognitive processes without the linguistic facility.

These examples provide some indication of the potential value of a somewhat altered context for the description and interpretation of biological processes. Since the viewpoint is developed in terms of concepts which apply most naturally to high-level behaviour, the context does not run the risk of a reductionist treatment of these levels. Furthermore, it makes certain consequences self evident. For example, the influence of an hypothesis or a model on the behaviour of an organism becomes axiomatic, since this is a basic determinant of the pattern which the organism follows in any given situation, from bacterium to man. Exactly how it enters as a determinant depends upon the level of organization one is considering, but in every one it is postulated as fundamental to the process observed. The hypothesis may be considered to specify certain basic constraints entering as contingent factors in the organic process, supplementing in a characteristically biological manner the 'laws of motion' of the organism seen as a physicochemical system. These constraints specify what the system can actually 'observe', how it can respond to its environment. Thus all organisms became observers conditioned by hypotheses. Whether or not this viewpoint can contribute anything constructive to the problem of observation and measurement in physics is far from clear ; but at the very least, the language is the same. This requires caution, since analogies can mislead ; but it could also lead to clarification.

However, I am more concerned here with the consequences of the breakdown of the concept of the objective observer in relation to the problem of meaning and wisdom, man's responsible participation in the world process. So long as

271

man was regarded as a simple observer of nature, he could not be held responsible for what he observed. He could be held responsible for the way he used any power resulting from his observations, but the truth he deduced therefrom was considered to be as inevitable a consequence of the observations as the deductions of Euclidean theorems from the axioms. But if one is responsible for the axioms, then one is responsible for the theorems. This is the present situation, truth being as relative to viewpoint as theorems are to axioms. For example, the statement that psychological states are 'nothing but' the behavioural consequences of hormonal change is a perfectly correct deduction from particular assumptions, and has an indisputable element of validity. But a total commitment to an essentially reductionist view of this kind involves a value judgement that needs examination. It is often associated with an essentially stoical attitude to the truth : for example, human beings are very subject to romantic illusions about who and what they are, the 'truth' being that they are 'nothing but' this or that. The value judgement is that it is good for people to face this 'reality', much as it is sometimes believed that good medicine must have a nasty taste.

It is often claimed in justification of a particular view or theory that it gives the best fit to the available data, and hence the criterion of choice is itself subject to objective measurement. When theories are in conflict, scientists ought therefore to choose the one which gives the most accurate predictions. However, it is quite clear from the history of science that this criterion is far from adequate to explain the choices that scientists have in fact made. As Kuhn has stated in his detailed study, *The Copernican Revolution* [11] :

Judged on purely practical grounds, Copernicus' new planetary system was a failure ; it was neither more accurate nor significantly simpler than its Ptolemaic predecessors. . . . as Copernicus himself realized, the real appeal of sun-centred astronomy was aesthetic rather than pragmatic.

It is now widely recognized that the factors determining which of two competing theories will survive are not in the least obvious, and some eminent scientists such as Dirac and Schrödinger have been quite explicit about the importance of elegance, simplicity, and beauty in theories. These are undefined terms which need examination, but it is evident that they refer to values, not to objective criteria of measurement. Once again we are forced to recognize the two-way nature of the dialogue with nature : we can understand only what is intelligible to us ; but at the same time, our questions must be intelligible to nature : they must, in some sense, be well formed to generate a response.

Because there is a measure of freedom which scientists can exercise in choos-

B.C. Goodwin .

ing a particular viewpoint or theory or paradigm, they are in the same measure
responsible for the viewpoints they adopt and should be prepared to justify them
on moral grounds. The image man has of himself and of the world is an exceedingly potent factor in conditioning his behaviour towards his fellow men and his
environment. We thus come full circle round to Alchemy again, but displaced one
step up the spiral staircase. The 'truth' we discover in the world is strongly conditioned by the image we hold of it (the world), our attitude towards it, as the
Alchemist believed. The difference lies in the fact that whereas the Alchemist
believed that one's attitude entered the natural process as a causal factor, we
now see it entering as an interpretative factor. Our viewpoint probably doesn't
alter the actual phenomenon observed (the 'probably' entering here to allow for
the possibility of para-psychological phenomena) : but it heavily conditions our
understanding of the phenomenon ; that is, into what context we choose to fit it,
to make it comprehensible. There is a context of understanding which involves
man's commitment to a particular form of participation in the world. This context
includes meaning as well as knowledge and can lead to wisdom as well as
power. The exact form of this context is not at all clear, and ultimately rests in
the hands of each individual.

▶ *Embryology and ethics.* Since I am professionally engaged in the study of
embryology the problem of ethical participation presents itself to me in this context. What can this mean in relation to activities such as vivisection, the mutilation of living organisms like Hydra or amphibian embryos, or the 'sacrifice' of
mice and rats for the study of growth control in mammalian tissues ? The usual
way of justifying such interference with the integrity of organisms is to say that
these studies are intended to lead ultimately to benefits in medical science, to the
alleviation of human suffering and the improved health of mankind. I believe that
to be both true and important, but it does not seem to me to be a sufficient
justification for embryological research.

In order to approach an answer to this problem, let me ask another question.
How can one justify the destruction of plants and animals to provide, say, wood
for houses or food for oneself and for others ? It is generally accepted that such
actions are justified so long as basic human needs only are satisfied. Wanton
destruction or slaughter is not condoned ; it is unethical. Behind this attitude is
the belief that humanity is part of a larger process of nature to which we should
contribute in our own way. Our existence is justified only if we cooperate with
this process, not if we violate it. Each person has his own answer to the question
what this process may be and how we fit into it. My own belief is that our

273

Biology and meaning

contribution is to live our lives in as full and integrated a manner as possible, and in so doing to help others to do so. One of our characteristic attributes as human beings is the possession of consciousness. Fulfilment of human potential is possible only if we develop this attribute to the greatest extent we are capable of. This includes the discovery of knowledge, but knowledge must be developed within a context of meaning, not simply for power. In so far as our actions are dedicated to the enhancement of consciousness for the purpose of achieving wholeness and completion, for healing, making wholes out of parts, our actions are ethical and there will be an experience of meaning in creative participation. Thus the destructive interference with the nucleus of the atom or the developing embryo is justified if the knowledge so obtained is used to assist in the completion of integrated, cooperative relationships within and between individuals and with the world. This is creative activity which for human beings is inextricably associated with ever-increasing consciousness, involving a greater and greater sense of responsibility as one becomes more and more aware of the ways in which parts are and can be united into ever more extensive wholes.

Perhaps the most striking fact about embryos and creatures with regenerative powers like Hydra is their extraordinary capacity to make wholes out of parts, to realize themselves as complete entities despite various disturbances. There are, of course, limits : they are unable to recover from too severe a disturbance. But within these limits the developing or regenerating organism undergoes transformations which produce ordered, harmonious, and balanced relationships between their cells, tissues, and organs. They do this by the combined processes of differentiation of elements and their cooperative union into the whole which gives meaning to the elements. This is a very remarkable spectacle and not only brings one into a relationship of understanding with the developmental process, but also provides a metaphor for human and social transformations. Here, too, it is necessary to understand the specific functions of parts and how they may be cooperatively united into wholes. This metaphor is, of course, a very old one, a good bit older than the Greeks. And metaphors can be very misleading, so one should be exceedingly careful to avoid a simple identification of one process with another. Psychical and social transformations are potentially unending, new dialectical tensions constantly arising from previous resolutions, unlike the embryo which reaches a terminal state in the adult form. The contrast here is the same as that between ontogenetically fixed genetic hypotheses and free, non-terminating mental hypothesis construction, as previously described. The significance of the metaphor is to be found in the vision that inspires it, which is that

274

man and Nature are distinct but united. The world is both intelligible and meaningful because we reflect its basic structures and participate in its processes,
much as the shape of a fish both reflects the hydrodynamic properties of the
water in which it lives and also allows it to participate in water movements.

These simple observations clearly do not give any prescriptions for judging
particular actions as ethical or unethical, since ethical choice becomes context-
dependent rather than universal. Nor do they provide any automatic criterion of
choice between competing scientific viewpoints or theories. However, there is
obviously a very important dimension of value which can enter consciously into
this choice so that science, far from necessarily destroying values, can itself
become inextricably connected with the conscious exercise of ethical decision.
Science could then become a transformed and transforming activity which
engages man totally in responsible creativity, bringing knowledge and power
into union with meaning and wisdom.

References

1. *Musaeum Hermeticum* (Frankfurt 1678).
Translated by A. E. Waite : *The Hermetic
Museum Restored and Enlarged* (London
1893, 2 volumes. Reprinted 1953).
2. C. G. Jung, *Psychology and Alchemy*, 2nd
edition, vol. 12 of (H. Read, ed.) *Collected
Works* (Routledge and Kegan Paul : London
1968).
3. C. G. Jung, *Mysterium Coniunctionis*,
vol. 14 of (H. Read, ed.) *Collected Works*
(Routledge and Kegan Paul : London 1963).
4. K. Seligmann, *History of Magic* (Pantheon
Books : 1948).
5. T. S. Kuhn, *The Structure of Scientific
Revolutions* (Chicago University Press 1962).
6. H. Kearney, *Science and Change 1500–
1700* (World University Library 1971).
7. A. Malraux, *The Voices of Silence* (Double-
day : New York 1953).
8. R. G. Collingwood, *The Idea of Nature*
(Oxford University Press 1945).
9. E. W. Bastin and C. W. Kilmister, *Proc. Camb.
Phil. Soc. 50* (1954) 278-91.
10. E. W. Bastin, On the origin of the scale
constants of physics. *Studia Philosophica
Gandensia 4* (1966) 77-101.
11. T. S. Kuhn, *The Copernican Revolution*
p. 171 (Harvard University Press 1966).

Appendix. A catastrophe machine

E. C. Zeeman
University of Warwick

This is a small toy made out of elastic bands designed to illustrate catastrophe theory, and in particular the cusp (or Riemann-Hugoniot) catastrophe [1-4]. I designed it for pedagogical reasons, in order to provide a concrete example in which all the variables were obvious and measurable, and the relationship between them differentiable and computable. But then I found it so illuminating to experiment with, giving such a powerful intuition of how a continuous change of control can cause discontinuous jumps in behaviour, that I strongly recommend the reader to make one for himself or herself.

Materials needed: Two elastic bands, three drawing-pins, a piece of cardboard and a piece of wood (the desk-top will do if you don't mind sticking drawing-pins into it).

Take as unit length the length of the unstretched elastic bands. With this unit the wood needs to be about 2 × 6. Cut a cardboard disk of unit diameter. Let X be the first drawing-pin. Use X to pin the centre of the disk to the centre of the wood, so that it spins freely. Make sure that it does spin freely by enlarging the hole if necessary—sometimes a small washer under the disk helps. Stick drawing-pin Y into the wood a distance 2 from X. Now fix the elastic bands to the disk at a point near the circumference; the easiest way to do this is to stick drawing-pin Z *upwards* through the disk near the circumference, tie the two elastic bands together in a reef knot, slip the knot over the point of Z and pull it tight. (Of course the knot does have a tendency to slip off occasionally, but the enthusiast no doubt can replace drawing-pin Z by a suitable nut and bolt.)

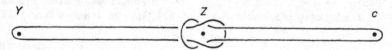

FIGURE 1

Now slip the other end of one of the elastic bands over Y, and the machine is complete, ready to go.

Hold the other end c of the other elastic band so that both elastics are taut, and move c smoothly and slowly about the plane. The disk will respond by moving smoothly and slowly most of the time, but occasionally it will make a sudden

Appendix

jump. This jump is called a *catastrophe*. Now put a pencil dot where c was each time the machine makes a jump. Very soon the dots will build up a concave diamond-shaped curve with cusps P, P', Q, Q' (see Figures 2 and 6). This curve is called the *bifurcation set B*. The curve is symmetrical about the line XY. The cusps P, P' lie on the line XY at distances approximately $XP = 1 \cdot 41$, $XP' = 2 \cdot 46$.

The reader will observe that the machine does not jump when c enters B, but only where c exits from B, and then only provided c has previously entered B from the opposite side. If c makes a complete circle round B, then the disk executes a smooth circle. If c makes little circles round P or P' then the disk hiccoughs once each time round. By contrast if c makes little circles round Q or Q' then the disk makes smooth oscillations. All this qualitative behaviour is surprising at first, but becomes easy to understand with the pictures of catastrophe theory. Therefore let us put the mathematics of the machine into the framework of catastrophe theory. Define

Control point $= c =$ the held end of the elastic.

Control space, $C =$ plane $=$ all possible positions of c.

State, $\theta =$ angle $Y\hat{X}Z =$ position of disk.

State space, $S =$ circle $=$ all possible positions of disk.

Potential $V_c(\theta) =$ potential energy in elastic bands with control held at c, and disk held at θ.

Therefore

$V_c : S \rightarrow R$ is a smooth function from the circle S into the reals R, and

$V : C \times S \rightarrow R$ is a smooth function $C \times S$ into the reals.

If c is held fixed and the disk is released then the disk will jump into a state θ_c, that is a local minimum of V_c, and stay there. Friction damps out any oscillations, so that the system is dissipative. Therefore the machine obeys the fundamental requirement of catastrophe theory, that the state θ seeks a local minimum θ_c of the potential V_c. It transpires that if c lies *outside* B then V_c has one minimum (and one maximum). Therefore the position of the disk is determined uniquely. However if c lies *inside* B then V_c has two minima (separated by two maxima). The choice of which of the two minima that potential V_c will take, and therefore which of the two positions the disk will seek, is determined by the past history of c, as follows. If c moves smoothly then a minimum of V_c will move smoothly, unless it happens to be annihilated by coalescing with a maximum as c exits from B. In this case, if the disk were in the minimum that was annihilated, then it will have to jump into the other minimum. This is illustrated by Figure 3 which shows a sequence of potentials V_c as c runs along the dotted line in Figure 2, near the

277

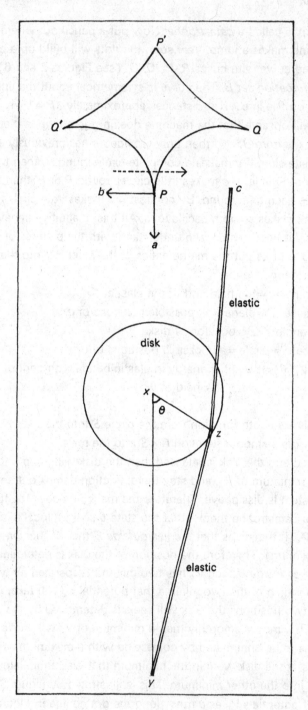

FIGURE 2

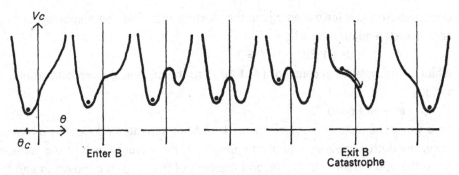

FIGURE 3

cusp P, parallel to the b-axis. In each case the graph is drawn for small value of θ.

The next diagram, Figure 4, shows the graph of θ_c as a function of c, drawn for small values of control coordinates $c = (a, b)$ with origin at the cusp P, and for small values of θ. More precisely the graph is given by $\frac{\partial V}{\partial \theta} = 0$, because the graph includes not only the minima of V_c but also the maxima of V_c near $\theta = 0$, shown shaded. The other single valued surface of maxima near $\theta = \pi$ is not shown in Figure 4 (but is shown in Figure 5).

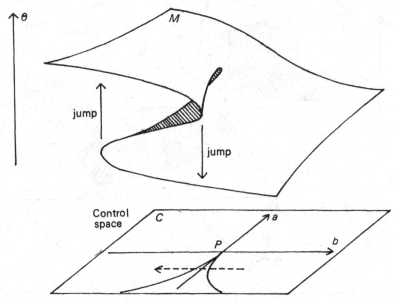

FIGURE 4

279

Appendix

Using Hooke's law for the energy in the elastic bands one can compute the approximate equation for M :

$$\theta^3 - 1 \cdot 3b\theta^2 + 1 \cdot 8a\theta + 1 \cdot 3b = 0 \quad,$$

which is equivalent (in sense of [4, 1.10]) to the canonical cusp catastrophe surface

$$\theta^3 + a\theta + b = 0 \quad.$$

As c varies the state θ_c stays on the surface M of minima. Therefore as c moves along the dotted line the state has to jump from the lower branch of the surface on to the upper branch at the second crossing of the cusp. If c moves to and fro the state performs hysteresis action.

We may summarize Figures 3 and 4 by saying that the machine obeys the *delay convention* which says 'delay jumping until necessary'. Delay is a consequence of the fact that the dynamics of the damped disk obeys a differential equation in θ with the property that θ and $\frac{\partial V}{\partial \theta}$ have opposite signs. Any appli-

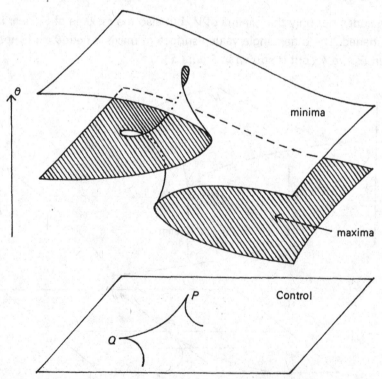

FIGURE 5

280

cation of catastrophe theory satisfying a differential equation with this property will exhibit delay—for example the applications to heartbeat and nerve impulse [4]. By contrast the application to phase-transition [1] does not exhibit delay. The reason is that the variables temperature, pressure and density in phase-transition do not obey a differential equation direct, but are averaging devices, exhibiting the average behaviour of very many differential equations obeyed by particles of the substrate.

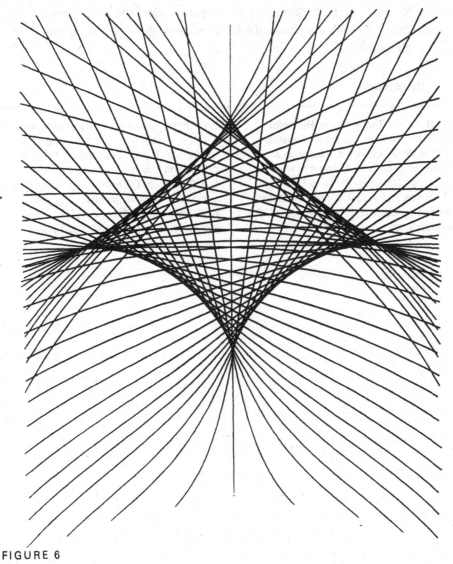

FIGURE 6

Appendix

If we now turn our attention to the cusp Q we find a surface similar to Figure 4, but dual in the sense that maxima and minima are interchanged. If we put the pictures for P and Q together we obtain Figure 5. It is difficult to combine pictures for all four cusps, because this would involve bending the θ-axis round into a circle. We leave this as an amusement for the reader.

The precise shape of the bifurcation set B has been studied by Poston and Woodcock [2]. They have drawn it accurately by computer, and I am grateful to them for allowing me to reproduce it in Figure 6. Here the lines represent the intersections of M with the planes $\theta =$ constant, and hence B is the envelope for all such lines.

References

1. D. H. Fowler, The Riemann–Hugoniot Catastrophe and van der Waals Equation, in (C. H. Waddington, ed.) *Towards a Theoretical Biology 4: Essays* pp. 1-7 (Edinburgh University Press 1972).
2. T. Poston and A. E. R. Woodcock, Zeeman's Catastrophe Machine, *Proc. Camb. Phil. Soc.* (In press).
3. R. Thom, Topological models in biology, in (C. H. Waddington, ed.) *Towards a Theoretical Biology 3: Drafts* pp. 89-116 (Edinburgh University Press 1970).
4. E. C. Zeeman, Differential equations for the heartbeat and nerve impulse, in (C. H. Waddington, ed.) *Towards a Theoretical Biology 4: Essays* pp. 8-67 (Edinburgh University Press 1972).

Epilogue

C. H. Waddington

University of Edinburgh

In introducing the first volume of this series, based on the symposium held in 1966, I wrote that: 'Theoretical Biology can hardly be said to exist yet as an academic discipline. There is even little agreement as to what topics it should deal with or in what manner it should proceed. . .' We set out 'to explore the possibility that the time is ripe to formulate some skeleton of concepts and methods around which Theoretical Biology can grow'. Perhaps one would now, four symposia later, ask what progress, if any, has been made Towards a Theoretical Biology.

I think the answer might begin by pointing out that the will-o'-the-wisp we have been pursuing should probably not have been labelled Theoretical Biology in the first place. A better name would have been Theory of General Biology. It has always been clear that we were not so deeply interested in the theory of any particular biological phenomenon for its own sake, but mainly in so far as it helps to a greater comprehension of the general character of the processes that go on in living as contrasted with non-living systems. This general character can, of course, scarcely be profitably discussed in complete abstraction, and these volumes, including the present one, contain many papers which deal in considerable detail with particular biological processes, such as the morphogenesis of pattern, the development of neuronal networks, evolutionary processes and several others; but these have been presented in a context, and have usually been given a form by their authors, which tends to bring out their relevance to the general underlying nature of living systems.

If one were to ask for a more definite description of what seems to have emerged about that general nature, probably most of the symposiasts would be tempted to reply by posing a series of questions rather than by making a series of statements; and probably also, each one would put rather different questions. I can here only sketch some of the considerations that appear most cogent to me, without pretending to offer any agreed consensus.

The focal point seems to me to be the confrontation of microstates and macrostates. The microstates of biological systems must eventually be describable in terms of molecules and energy, since that is how we describe all microstates in material systems. If one allows the term 'molecule' to be used in a rather broad sense, to include the relatively huge macromolecular assemblies

283

within cells, one can agree that 'molecular biology' is making rapid and effective progress in improving our descriptions of the entities involved and of their causal interactions. None of the symposiasts seems to have felt much inclination to suggest that the crucial *questions* for a general theory of biology lie in that area.

The challenging and intriguing problems arise from the great difficulty—conceivably amounting to theoretical impossibility—of accounting for some of the most interesting phenomena of life in terms of molecular microstates. Perhaps the clearest statement of the problem in this volume is Kauffman's discussion of the way in which Elsasser has posed the matter. If a eukaryotic cell contains 10^5 or 10^6 genes, of which a large proportion are active at any given time; if a neuronal network is built of a similarly large number of cells, each in contact with some hundreds of others; how can we hope to analyze the microstates in sufficient detail to yield explanations of cellular differentiation or neural information processing? As Pattee puts it in his essay (p. 225), we find little difficulty in seeing how complexity can arise from simplicity, as in the self-assembly of more and more complex molecules and molecular assemblages; but we find ourselves having to understand how things with a certain global simplicity—a transfer of gaze, a tissue cell of a certain type, a four- or five-jointed leg—rise from a set of microstates of extreme complexity. Many of our discussions have therefore been explorations of ways of handling simple macrostates without having to break them down into an unmanageable plethora of vastly complex microstates.

One of the most fully developed approaches has been by way of René Thom's theory of catastrophes. This is a general theory about discontinuities, which may divide a multivariate phase-space into regions which have each a definite identity, since they are bounded against each other by the catastrophe surfaces. We have here a conceptual framework which is not atomistic, but which provides clear-cut boundaries and precludes interpenetration of entities, features which are two of the major aids to clarity of thought which atomism has provided in the past. Fowler's, and particularly Zeeman's, essays, show how, in the latter's words, such thinking allows us 'to model the dynamics (which is relatively simple) rather than the biochemistry (which is relatively complicated)'.

Kauffman's studies on switching nets, described most fully in Volume 3, but touched on in his essay here, are in a way the obverse of the same point. Setting up networks with connections which look as though they should guarantee chaotic complexity, he shows that global simplicity of macro-behaviour in fact emerges surprisingly often. Another example is provided by Elsdale's Inherently Precise Mechanism, by which cells randomly spread on a flat surface assemble themselves into arrays which have a certain, though relatively low, degree of simplicity. But Elsdale's observations bring up the next step which needs to be taken along this line of thought; the partly assembled arrays (macrostates) influence the behaviour of the individual cells (microstates), for instance by letting them fall into the cracks if they are oriented nearly parallel to the array, or supporting them if they are transverse to it (p. 100).

The situations which arise when there is mutual interaction between the complexity-out-of-simplicity (self-assembly), and simplicity-out-of-complexity (self-organization) processes, are, I think, to be discussed most profoundly at the present time with the help of the analogy of language. This has been pursued, in the present volume, by both Thom and Pattee. Pattee emphasizes that a language must clearly be different from the subjects which can be talked about in it; a symbol can only be such, if it is not identical with what is symbolized. Further, we decide which is the symbol and which the symbolized by the distribution of our general interest, which is attracted to the latter: we are in general more interested in cats than in the word-symbol *cat*, otherwise we would take the word as the thing and the furry animal as the symbol.

Now the living world, as Pattee points out, is founded on a dualism of a kind which can be regarded in this way. At a relatively simple level, we have the genotype, of relatively unreactive DNA, in which no one could work up much interest were it not that it can act as a symbol coding for the much more reactive phenotype, the proteins which perform operations on their surroundings. It is on the phenotypes that our immediate interest is directed as natural predatory members of the biosphere. So is that of natural selection; a large section of Volume 2 was devoted to discussion of this point in an evolutionary context between myself and Maynard Smith.

Returning to Pattee's essay; he argues that a symbol can only function as such when it is part of a system of symbols. A word must be a word in a language. To his own question 'How does a molecule become a message' he answers that it is inadequate merely to say 'when it codes for some other molecule in which we are more interested', as a nucleic acid may code for a protein; such a reply tacitly assumes the existence of ribosomes, polymerases, activated amino-acids and so on—a whole linguistic apparatus, a grammar. The 'structures mediating global simplicity' which we have to search for in the theory of general biology are, then, perhaps profitably to be compared with languages; based on

Epilogue

the primary biological disjunction between genotype and phenotype as the analogue of symbol-symbolized, but going as much beyond it as a structured grammar is beyond the single word. And once we have a language, we can have a metalanguage; we can define one word in terms of a set of other words.

Most of the biological problems we have discussed can be seen in terms of the language-metalanguage analogy. I have already mentioned the cell bundles of Elsdale's experiments which exert a 'meta-linguistic' influence on still-isolated individual cells. In my own discussion of patterns in Drosophila, I was concerned to show how, from the extreme complexity of the many-gened genotype, relatively simple 'linguistic' structures may be formed, such as the expanding column of cells which is the rudiment of the tarsus; and a slight change in one of the 'words' in this (substitution of the mutant for the normal allele at the *four-jointed* locus, gives a 'sentence' with the same structure but an altered meaning—a four- instead of a five-jointed leg (p. 133). Or, in a wing whose 'grammatical structure' has been altered by the mutant genes *approximated* or *shifted*, there is a 'metalinguistic' effect on the position of the cross-vein (p. 131). My argument with Wolpert on p. 145 could be expressed by saying that I feel that positional information is only a metalanguage of pattern, and that he is neglecting the underlying language by lumping it into an unanalyzed 'cellular competence'. Finally, both Arbib and Cowan are concerned with ways in which exceedingly complex assemblages of neurones give rise, by an inter-play between self-assembly (formation of inter-connections) and self-organization (previous connections influencing how later connections are made) into coherent structures complex enough to be said to have a 'grammar' and to function as 'languages'.

Theorists of general biology are only just feeling their way towards the language-metalanguage analogy which I have expressed, rather brashly because so summarily, in the last few paragraphs. It remains to be seen how valuable will be the insights which it can provide. At any rate biologists who pursue this line of thought will be assured that they will find themselves in stimulating company. The study of the development and fundamental nature of language is one of the most active fields of thought at the present time; not only where one would expect to find it, among students of human behaviour, such as Chomsky, Harris, Piaget and others, but by computer scientists such as Longuet-Higgins and Winograd. Perhaps I should have foretold, when I wrote the introductory essay to the first volume, and stressed that biology is concerned with algorithm and programme, that this would be the company in which we should eventually

286

find ourselves. But, although I was pretty sure that we should not end up amidst the niceties of negentropy and Information Theory, the aridities of Neo-Darwinism algebra, or even the fraction collectors and Spincos of molecular biology, I did not at all clearly see the direction in which we did actually move. I doubt if many others did either. Even if we turn out to have taken a wrong course, at least this progression may be taken as evidence that the four Symposia at the Villa Serbelloni have not been totally chaotic, but have taken a few steps in one rather definite and not too hackneyed direction, which I am sure will be interesting and think will be useful.

A fine peroration, I hope you will think. But I suppose most of my readers will be scientists, and one of the most valuable characteristics of scientists is to be dirty-minded, to ask awkward questions. 'What', you may well ask, 'is this thing, a language, you refer to? It has of course been a fashionable court of last appeal, from Karnap through the developmental history of Wittgenstein to Chomsky and beyond. These are elementary points you've made, that word-symbols must be different from things-symbolized, and that any one word-symbol is vacuous except in the context of a language. But what is a language essentially?'

In my opinion, biology can make some real contribution towards answering this question. Conventionally, among language theorists from Wittgenstein (early—and even possibly late) to Chomsky, language is discussed as though its primary concern is with making statements. 'The flower is red.' But who cares whether it is red or blue? To the biologist, language is something which must have evolved. And, as I have insisted so many times before in these volumes, the driving force of evolution, natural selection, is concerned with *effects*. Mere statements, such as the sequence of nucleotides in DNA which describe the arrangement of amino acids in a protein, fall outside the category of things it can get its hands on to, which are relevant to it. Natural selection is concerned with actions in relation to the existing circumstances; what the proteins *do*, as enzymes or what have you.

Human language must have been produced by evolution. Evolutionary forces —and that means natural selection, the only evolutionary force we know of— could have produced it only if it produced effects. And a coding of natural phenomena into symbols, into a language, could only produce effects if the basic character of language is to be imperative, not indicative; to express, in symbolic form, commands, instructions, programmes—not statements.

The fundamental form of 'generative grammar' on this view is not: Noun

287

Epilogue

Phrase—Verb Phrase. It is: *You→Do*. It must be the Second Person; if it were the First, *I* do, there is no point in saying anything; if it is the Third, *He* do, it becomes again a mere description, a statement. And that would conflict with the essential necessity for the *'Do'*—if that is omitted, no effect is produced, and natural selection can take no interest.

This is, of course, the grossest heresy against the orthodoxy with which people have attempted to indoctrinate me since I was a young man. The whole gist of language, I was told, is concerned with statements—analytic or empirical. A command cannot be asserted or denied, and, according to Karnap for instance, is therefore 'without sense'. Present-day linguistic philosophers have relented to some extent from the extreme Fundamentalism of Logical Positivism, but, in my view, not nearly enough. To the natural selective forces which bring about evolution—of language, as of anything else—it is precisely commands, producing effects, which do have sense, and it is statements about facts—atomic or otherwise—which are without significance, or meaning.

In so far as a human child has at birth any genetically determined 'competence for acquitting language', this should not be regarded as a special faculty unrelated to anything else in the arsenal of skills gradually maturing in the neuronal networks of his brain. It must be a part—a very sophisticated part—of his intentional motor skills, by which he *affects* his surroundings, and scores at some point on the scale of 'natural selective value'. And if anyone tells me—I have not checked up recently—that Piaget or Bruner or someone, has shown that young children can handle language about activities which their motor control does not yet allow them to perform—well, I have a whole biological theory of genetic assimilation and epigenetic anticipation which could cope with this; and foetuses have lungs well designed for transferring atmospheric oxygen to the blood long before the atmosphere impinges on them. Don't let's be diverted from the fundamental and theoretical points about the nature of language by empirical considerations about how far the genetic assimilation of language competence has gone in human evolution. It would not really contribute to our understanding of the nature of language and its relations to the 'real world', if it had happened to be the case that every newborn babe could speak fluent Hebrew —it would have to be the language of the Garden of Eden, I suppose.

To a biologist, therefore, a language is a set of symbols, organized by some sort of generative grammar, which makes possible the conveyance of (more or less) precise commands for action to produce effects on the surroundings of the emitting and the recipient entities. It is, I think, significant that the people at the

288

Epilogue

Serbelloni meetings who have been concerned to analyze language into a form in which computers can participate in its use (e.g. Longuet-Higgins), have found themselves forced to a view very like this — that the basic sentences in language are programmes, not statements. And it is language in this sense — not as a mere vehicle of vacuous information — that I suggest may become a paradigm for the theory of General Biology.

List of Participants

Michael **Arbib** — Automata Theorist, University of Massachusetts, Amherst.

Morrel H. **Cohen** — Physicist and Biologist, Director, The James Franck Institute, University of Chicago.

Michael **Conrad** — Theoretical Biologist, Department of Mathematics, University of California.

Jack **Cowan** — Mathematical Neurobiology, Department of Mathematical Biology, University of Chicago.

Francis **Crick** — Molecular Biologist, MRC Laboratory of Molecular Biology, Cambridge

Thomas **Elsdale** — Cell Biologist, MRC Cytogenetics Research Unit, Western General Hospital, Edinburgh.

David **Fowler** — Mathematician, University of Warwick.

Brian **Goodwin** — Theoretical Biologist, The University of Sussex.

Richard **Gordon** — Theoretical Biologist, Center for Theoretical Biology, State University of N.Y. at Buffalo.

Stuart A. **Kauffman** — Theoretical Biologist, Committee on Mathematical Biology, Chicago.

C. **Longuet-Higgins** — Professor of Machine Intelligence, University of Edinburgh.

Howard **Pattee** — Physicist, Department of Theoretical Biology, State University of N.Y. at Buffalo.

René **Thom** — Topologist, Institut des Hautes Études Scientifiques, France.

J. P. **Thorne** — Linguist, Department of English Language, University of Edinburgh.

C. H. **Waddington** — Biologist, Institute of Animal Genetics, University of Edinburgh.

Lewis **Wolpert** — Developmental Biologist, Department of Biology, The Middlesex Hospital Medical School, London.

Christopher **Zeeman** — Mathematician, University of Warwick.

Author Index

Bold type indicates beginning of article

Author Index

Subject Index

Subject Index

Subject Index

Subject Index

Subject Index

Printed in the United States
by Baker & Taylor Publisher Services